FIELD SOIL WATER REGIME

DuP?

Jou Spe

FIELD SOIL WATER REGIME

Proceedings of a symposium held on August 16, 1971, during the annual meeting of the Soil Science Society of America and American Society of Agronomy. Sponsored by Divisions S-5, A-3, S-1, S-6, and S-7 of the societies.

Editorial Committee

R. RUSSELL BRUCE, Chairman

*Supervisory Soil Scientist, USDA Agricultural Research Service,
Southern Piedmont Conservation Research Center,
Watkinsville, Georgia*

KLAUS W. FLACH

*Director, Soil Survey Investigations, USDA Soil Conservation Service,
Washington, D. C.*

HOWARD M. TAYLOR

*Research Soil Scientist, USDA Agricultural Research Service,
Iowa State University, Ames, Iowa*

Coordinating Editor
MATTHIAS STELLY

Managing Editor
RICHARD C. DINAUER

Assistant Editor
JOYCE M. HACH

Number 5 in the
SSSA SPECIAL PUBLICATION SERIES

Soil Science Society of America, Inc., publisher

Madison, Wisconsin USA

1973

The cover design is by Garnet Schuch, former
staff artist with the American Society of
Agronomy, Madison, Wisconsin

Soil Science Society of America, Inc.
677 South Segoe Road, Madison, Wisconsin 53711 USA

Library of Congress Catalog Card Number: 73-87811

Printed in the United States of America

CONTENTS

4   **Experiments in Predicting Evapotranspiration by Simulation
     With a Soil-Plant-Atmosphere Model (SPAM)**

E. R. LEMON, D. W. STEWART, R. W. SHAWCROFT and
S. E. JENSEN

5   **Relationships Between Soil Structure Characteristics and
     Hydraulic Conductivity**

J. BOUMA and J. L. ANDERSON

6   **Water Retention and Flow in Layered Soil Profiles**

D. E. MILLER

7 Field Moisture Regimes and Morphology of Some Arid-Land
Soils in New Mexico

CARLTON H. HERBEL and LELAND H. GILE

8 Oxygen Content in the Ground Water of Some North Carolina
Aquults and Udults

R. B. DANIELS, E. E. GAMBLE, and S. W. BUOL

9 Hydrology and Soil Science

C. R. AMERMAN

10   The Role of Soil Water in the Hydrologic Behavior of Upland Basins

WADE L. NUTTER

11   Development of Soil Temperature and Soil Water Criteria for Characterizing Soil Climates in Canada

WOLFGANG BAIER and ALEX R. MACK

FOREWORD

There is no doubt that water plays a significant role in determining the nature and properties of soils. It dissolves soil minerals and is the means whereby the products of dissolution are transported from one horizon to another. It governs the permeation of air into soil pores and, hence, chemical and biological oxidation and reduction. By reason of its relatively large heat capacity and heat of fusion, it has a profound influence on soil temperature. As it filters into or evaporates from the soil, the attendant swelling and shrinking affect soil structure. Its flow over and into the soil redistributes the soil particles. In its absence, life in the soil comes to a virtual standstill. Indeed, all the physical, chemical, and biological reactions and processes that occur in the soil are dependent on its presence.

Water in the soil is absorbed, utilized, and transpired by the plants growing on it. Further, as already noted, it affects the properties of the soil and the reactions and processes that occur therein. Plants are sensitive to these properties, reactions, and processes. Hence, soil water affects plant growth and development both directly and indirectly.

In view of the importance of soil-water interrelations and their quantitative description, a symposium was held in 1971 at the annual meetings of the Soil Science Society of America on the topic "Field Soil Water Regime." This volume includes the papers presented at that symposium. Although it is recognized that these papers do not include all aspects of the subject, it is hoped that they will facilitate understanding and illustrate the wide range of activities related thereto.

August 1973

PHILIP F. LOW, *president*
Soil Science Society of America

PREFACE

In recent years quantitative description of the field soil water regime has become increasingly common. There are perhaps three main reasons. One is that physical theory and the required laboratory verification now would seem adequate to describe the water regime of soils under field conditions. The second is that it is now possible by computer to make calculations easily that involve the interactions of several variables; previously such calculations were very laborious or impossible. And thirdly, there is the emphasis in recent years on activities designed to better understand the soil as a medium for the disposal of pollutants. This concern for environmental questions has increasingly led experimentalists to perform experiments outdoors. This same concern for environmental questions has led soil morphologists and others concerned with description of natural conditions to become more concerned with the numerical description of the soil water regime.

The time, therefore, seemed right to present a sample of the kinds of research within the Soil Science Society of America and the American Society of Agronomy on the description of the soil water regime. Speakers and subjects were selected to provide a broad spectrum of approaches and research results. An attempt was made to include both papers that would orient the reader generally in a subject area and papers that are highly specific to a particular research interest. Some of the papers are highly quantitative; others are largely descriptive of the natural condition. Taken together, the papers illustrate the wide range in activities designed to characterize the soil water regime. This was the intent in their selection.

This group of papers is not intended to provide a coherent review of the subject. Rather, it is hoped that they would provide windows from one field of specialization to another. In this way they may help to increase the coherence of the subject area for the future.

September 1972

ROBERT B. GROSSMAN,
symposium chairman and head,
Soil Survey Laboratory
Soil Conservation Service, USDA
Lincoln, Nebraska

Conversion Factors for English and Metric Units and Plant Nutrients

To convert column 1 into column 2, multiply by	Column 1	Column 2	To convert column 2 into column 1, multiply by
	LENGTH		
0.621	kilometer, km	mile, mi	1.609
1.094	meter, m	yard, yd	0.914
0.394	centimeter, cm	inch, in	2.54
	AREA		
0.386	kilometer^2, km^2	mile^2, mi^2	2.590
247.1	kilometer^2, km^2	acre, acre	0.00405
2.471	hectare, ha	acre, acre	0.405
	VOLUME		
0.00973	meter^3, m^3	acre-inch	102.8
3.532	hectoliter, hl	cubic foot, ft^3	0.2832
2.838	hectoliter, hl	bushel, bu	0.352
0.0284	liter	bushel, bu	35.24
1.057	liter	quart (liquid), qt	0.946
	MASS		
1.102	ton (metric)	ton (English)	0.9072
2.205	quintal, q	hundredweight, cwt (short)	0.454
2.205	kilogram, kg	pound, lb	0.454
0.035	gram, g	ounce (avdp), oz	28.35
	PRESSURE		
14.50	bar	lb/inch^2, psi	0.06895
0.9869	bar	atmosphere,* atm	1.013
0.9678	kg (weight)/cm^2	atmosphere,* atm	1.033
14.22	kg (weight)/cm^2	lb/inch^2, psi	0.07031
14.70	atmosphere,* atm	lb/inch^2, psi	0.06805
	YIELD OR RATE		
0.446	ton (metric)/hectare	ton (English)/acre	2.240
0.891	kg/ha	lb/acre	1.12
0.891	quintal/hectare	hundredweight/acre	1.12
1.15	hectoliter/ha, hl/ha	bu/acre	0.87
	TEMPERATURE		
$\left(\frac{9}{5}\,°C\right) + 32$	Celsius	Fahrenheit	$\frac{5}{9}(°F - 32)$
	-17.8 C	0 F	
	0 C	32 F	
	20 C	68 F	
	100 C	212 F	

PLANT NUTRITION CONVERSION--P AND K

P (phosphorus) $\times$ 2.29 = P_2O_5

K (potassium) $\times$ 1.20 = K_2O

* The size of an "atmosphere" may be specified in either metric or English units.

Soil Moisture Regimes and Their Use in Soil Taxonomies[1]

GUY D. SMITH[2]

ABSTRACT

Wet soils have been distinguished from well-drained ones since men began to farm. Since V. V. Dokuchaiev pointed out that soils are natural bodies and can be classified as such, soil taxonomists have made the same distinction. Similarly taxonomists have distinguished what G. N. Coffey called "arid soils" from others. But most taxonomists have used accessory properties not the moisture per se to make the distinctions. Since 1960, there has been a slowly growing trend to use the soil moisture regime itself to define taxa.

The presence or absence of shallow ground water in soils has for centuries been considered an important soil property by those who till the soil, probably since men began to farm. The agric horizon in Belgium (R. Tavernier, personal communication) exists only in soils brought into cultivation at an early date and only in well-drained soils. Centuries ago, the cultivators obviously distinguished between well-drained and moderately well-drained soils. Hilgard (1860), writing about the soils of Mississippi, recognized the effects of ground water on soil colors, on concretions, on trafficability, and on crops. He implied that this was common knowledge. Soil scientists must have been aware of these relations ever since there were soil scientists, but they have been slow to use the presence or absence of shallow ground water as a soil property that defines kinds of soil, perhaps because artificial drainage can change the depth of ground water.

Nearly every classification that treats soils as natural bodies, beginning with that of Dokuchaiev (Sibertzev, 1900), has distinguished well-drained soils from poorly drained ones. Yet the most careful reading of the definitions written prior to about 1950 rarely shows whether the definitions were based on the water itself, or on properties that are associated with the water, or on the combination of these factors.

The soil scientists, not the cultivators, noticed that soils in arid regions differed in important respects from soils in humid regions. Few cultivators had the opportunity to see both arid and humid regions, or they too would have marked the differences. Dokuchaiev in Russia (Sibertzev, 1900) and Hilgard (1912) in the United States at about the same time were emphasizing the importance of *climate* to the nature of the soil. The moisture regime of the soil is related to the climate but the relation is imperfect because, as near-

[1] Contribution from the Soil Conservation Service, U. S. Department of Agriculture, Washington, D. C.

[2] Director, Soil Survey Investigations, now retired.

ly every cultivator knows, on a given farm some soils are wetter than others. Water runs off some soils and onto others.

Climate was recognized by both Dokuchaiev and Hilgard as one of the factors that determine the nature of the soil but they mentioned *climate,* not *soil climate.*

At about this same time, John Stuart Mill (1891, p. 499) wrote, "The properties, therefore, according to which objects are classified, should, if possible, be those which are causes of many other properties; or at any rate, which are sure marks of them. Causes are preferable, both as being the surest and most direct of marks, and as being themselves the properties on which it is of most use that our attention should be strongly fixed." Therefore, it might seem reasonable to have used soil moisture, which is the cause of many other properties, as a property to classify soils. But the early classifications of Dokuchaiev (1886) and Sibertzev (1900) did not do so. Hilgard (1912) did not attempt a classification although he wrote of the difference between arid and humid soils. Ramann (1911, p. 578 et seq.) and Coffey (1912) proposed classifications that included classes of arid soils. Their definitions, though relatively explicit, did not mention the moisture itself.

Glinka (1914) apparently was the first to propose the use of soil moisture to classify soils, although he says, "While we use moisture in this way as the most important basis for soil classification, we do not claim originality for the idea, but, on the other hand, we merely use the observations and conclusions of Russian and West European investigators." Glinka (1914) proposed that Ektodynamomorphic soils be grouped in six classes according to their moisture content, as follows:

Soils of optimum moisture content
Soils of average moisture content
Soils of moderate moisture content
Soils of insufficient moisture content
Soils of excessive moisture content
Soils of temporarily excessive moisture content

Distinctions between the classes were as vague as the names. Glinka's book *The Great Soil Groups of the World and Their Development,* translated from the German by Marbut, had an enormous effect in the United States.

Glinka (1931, p. 332 et seq.) in 1914 had proposed the use of moisture to classify soils, but he dropped the idea in 1921 and proposed that the Ectodynamomorphic soils be grouped into the following types:

I Lateritic
II Podzolic
III Steppe
IV Marshy
V Solonetz

It is difficult to be sure from reading Glinka's definitions (1931, p. 332 et seq.) whether the presence of shallow ground water was essential to the marshy type. Its presence is certainly implied but nowhere is it specifically stated to be required. Marbut did not accept soil moisture as the criterion for classifying soils. Rather, he looked for accessory properties. He recognized that the absence of shallow ground water was essential in what he considered the "mature soil." He says explicitly (1928b, p. 43), "The differentiation (into Pedocals and Pedalfers) *is not based on rainfall* (Marbut's italics), but the characteristics on which it is based are *explained* by different amounts of rainfall corresponding to the different characteristics." His report (1928a) on the First International Congress of Soil Science impresses on the reader the idea that climate is not a soil property and that soil moisture differences related to climate likewise are not soil properties that can be used to define kinds of soil. He said, "It is now clearly recognized that the so-called climatic groups are in fact character groups and that the classification, as far as it goes, is in reality based on characteristics, and that the nomenclature only is climatic." In none of the classifications of soil proposed between 1927 and 1947 was the moisture regime used to define taxa although definitions were ambiguous about ground water. However, in the midwestern USA by about 1940, soils that had relict gray colors and mottles, but that no longer had ground water, were being classified with moderately well- and well-drained Prairie soils (Brunizems) rather than with the Wiesenboden.

Kubiena (1953) defined classes of his highest category on the basis of presence or absence and continuity of ground water. However, it is more difficult to say whether he used moisture held at tensions greater than 0 to distinguish between classes. He describes some soils, such as the Alpine Pitch Rendzina, as being nearly always moist but never waterlogged, but this seems descriptive rather than definitive. Kubiena distinguished between soils with moisture rising from depth, with moisture moving downward in the soil, and with stagnant ground water in his classification, but statements about the duration and season of dryness in the soil seem to be largely descriptive.

Charles E. Kellogg has emphasized in conversations that soil temperature and soil moisture are soil properties that can be used to define taxa. As a result, the series of approximations of a new system of soil classification which began in 1951 included soil moisture regimes. They were tested and revised throughout all subsequent approximations.

The 7th Approximation (Soil Survey Staff, 1960) distinguished aquic suborders from others on the basis of colors combined with either saturation with water or with artificial drainage. It also laid the foundation for the present udic, ustic, and aridic moisture regimes. The xeric regime was introduced in the 1964 Supplement. Consistent usage of the moisture regimes was not reached until the 1967 Supplement was issued (Soil Survey Staff, 1967). Very briefly, the regimes proposed in the 1967 Supplement are now defined as follows (Definitions are simplified; for complete definitions, see Soil Survey Staff, 1973):

The moisture control section has an upper boundary at the depth to which a dry soil is moistened by 2.5 cm of water in 24 hours and a lower boundary at the depth to which the soil is moistened by 7.5 cm of water in

48 hours. Dry soil has water held at tensions $\geqslant$ 15-bar. Moist soil has water held at tensions $<$ 15-bar.

Aquic—In most years, the soil is saturated with oxygen-depleted water (tension is 0) at some season of biologic activity, but the water table may drop at another season.

Peraquic—The soil is always saturated as in the aquic regime, largely in tidal marshes.

Udic—In most years, the moisture control section is never dry in any part for as long as 90 days, cumulative, but is not aquic.

Perudic—Like udic but the soil is at or very near field capacity throughout the year.

Ustic—In most years, the moisture control section is dry in some part $>$ 90 days, cumulative, but is dry less than half the growing season in all parts in midlatitudes. In low latitudes, in most years, the moisture control section is moist in some part for more than 90 consecutive days but may be dry in all parts $>$ 90 days, or even $>$ 180 days, cumulative.

Xeric—In 6 or more years out of 10, the moisture control section is continuously moist $>$ 45 days in all parts in winter and is continuously dry in all parts in summer $>$ 45 days, and the soil temperature is thermic, mesic, or frigid. (Mean annual soil temperature is $<$ 22C.) The xeric regime is restricted to midlatitudes by a requirement that the mean winter soil temperature be $\geqslant$ 5C colder than that of summer.

Aridic and Torric—In most years, the moisture control section is dry in all parts more than half the growing season and is not moist in some part for as long as 90 consecutive days during a growing season.

It is recognized that few observations have been made on duration of dryness using the above definitions. But, there is a great deal of common knowledge among those who live in any area. At the moment, soils must be classified on the basis of the common knowledge. Should the classification of many be proven wrong, the definitions are more apt to be changed than the classification.

The uses one wants to make of a taxonomy have an important influence on the taxonomy itself. Soil moisture regimes are causes of many soil properties important to the use of the soil. They are critical to interpretations about soil-plant relations. Climatic phases, within limits, could substitute for soil moisture. But, as Herbel and Gile have pointed out (Chapter 7, this book), soils that lie side by side in the same climate may have different moisture supplies and may produce differing amounts of forage. Climatic phases cannot show such differences.

As an example, we might consider a soil formed in loess in western Iowa and another in the Palouse region of eastern Washington. Both soils had a grass vegetation, are nearly neutral, and are free of carbonates to considerable depths. Both were considered Brunizems in the classification used prior to 1965 because they had little or no secondary carbonate. There is such a complete overlap in the soils in the two regions in color, texture, reaction, structure, bulk density, thicknesses of horizons, mineralogy, soil temperature, and so on, that it would seem impossible not to have had some of the same series in both regions if we had had a rigorously defined national system of classification.

We might select a soil in the Palouse region that has a total summer precipitation of $<$ 65 mm out of an annual precipitation of 530 mm. It produces excellent winter wheat (*Triticum* spp.) with consistent high yields, but it will not produce corn (*Zea mays* L.) or soybeans (*Glycine max* L.) without irrigation. The corresponding soil in Iowa has a total summer precipitation of 330 mm out of an annual precipitation of 785 mm. It produces excellent corn, soybeans, and could also produce wheat. If these soils were in the same series, as they should be if soil moisture is not a differentia, it becomes difficult to interpret the soil maps. Complicated climatic phases are needed in addition to other phases, but there is no climatic classification that has been agreed upon even for the United States, let alone for North America or the world. Relations are very complex and no terminology is available. As a minimum, the climatic phase must specify summer and winter temperatures or degree days, length of growing season, and precipitation by seasons relative to evapotranspiration. This is in addition to the more conventional phases.

The problems of phasing for climate become even more complicated with Entisols, which can have any soil climate or external climate. The use of the soil climate in the taxonomy does not completely eliminate the need for climatic phases but makes the problems more manageable. Phases for length of growing season or degree days, for example, are still needed for interpretations.

The use of what Marbut called "character groups" is commonly satisfactory for small scale soil maps of plains but can cause serious troubles with interpretations on an individual farm or in mountainous areas. A soil that is not extensive in a state may be dominant on a single farm. The interpretations of the soils on a farm must be quite specific.

Rauzi is a tentative soil series that is extensive only locally in northern Wyoming. Because the parent materials are acid and there is little dust to bring in carbonates or salts, the Rauzi soils have no ca or sa horizon. They do have an argillic horizon. On the basis of these "characters", the Rauzi soils would be grouped with the Hapludalfs of Indiana and Ohio (Miami family) rather than with the surrounding Aridisols. Precipitation on the Rauzi soils is about 350 mm, with a maximum in the spring. Vegetation is short grass, primarily for grazing. Probably no one would be satisfied with a soil family that contained both the Miami and Rauzi series. But if we are to classify soils on their own properties, there seems no reasonable way to keep these soils apart other than by their differing moisture regimes.

The idea that moisture regimes are soil properties was not immediately accepted. At an international discussion (in 1952) by representatives of seven European national soil surveys and that of the United States, a modification of the 2nd Approximation was discussed. There was strong opposition expressed to the use of the soil climate, soil moisture, and temperature in a taxonomic classification.

The idea of using moisture regimes to define taxa is still not universally accepted by any means. Soil institutes responsible for small areas or for areas where the variability in soil moisture regimes is small, commonly are not concerned. Australian literature (Hallsworth, 1968) indicates that their soil scientists have little interest or are opposed. Soviet publications (Rozov & Ivanova, 1967) contain suggestions that the soil types can be grouped by temperature and moisture regimes, but there seems to be little agreement in the USSR as to how their soil types should be grouped. The legend of the FAO-UNESCO soil map of the world (Dudal, 1968, 1970), 1:5,000,000, recognizes soil moisture but not soil temperature as differentiae.

In the last decade, use of soil moisture has been increasing. In 1962, the New Zealand Soil Survey (Taylor & Pohlen, 1962, p. 157 et seq.) published a classification of New Zealand soils using moisture and temperature in definitions of classes.

In 1965 (Dudal, Tavernier, & Osmond, 1965), the soil map of Europe, 1:2,500,000, distinguished Gray-Brown Podzolic soils from Brown Mediterranean soils by the dryness of the latter soils in summer.

In 1967, the classification of soils for the soil survey of France (Aubert et al., 1967) distinguishes the presence or absence of ground water as well as its depth, and the dryness of the soil according to whether it is dry in winter or summer.

The Canadian soil survey in 1970 (Ehrlich), adopted the use of soil moisture classes as well as soil temperature classes. Variability of moisture is less in Canada than in the soils of the United States, but the classes are more narrowly defined to make them useful at the soil family level. In addition, the Gleysolic order is defined as being saturated and under reducing conditions part or all of the year, or the soils have artificial drainage.

The use of moisture regimes as differentiae is hampered by the scarcity of hard data on the seasonal changes in soil moisture, expressed in terms of energy. However, these data will be accumulated in the years to come, and definitions will be adjusted to group the soils into the most useful classes.

LITERATURE CITED

Aubert, G. et al. 1967. Classification des sols. Edition 1967. Commision de Pédologie et de cartographie des sols. Laboratoire de Géologie—Pédologie de L'E.N.S.A. Grignon.

Coffey, G. N. 1912. A study of the soils of the United States. U. S. Dep. Agr., Washington, D. C. Bureau of Soils Bull. 85.

Dokuchaiev, V. V. 1886. Cited by Sibertzev (1900).

Dudal, R. 1968. Definitions of soil units for the soil map of the world. World Soil Resources Report 33. FAO, UNESCO, Rome.

Dudal, R. 1970. Key to soil units for the soil map of the world. FAO, Rome.

Dudal, R., R. Tavernier, and D. Osmond. 1965. Soil map of Europe 1:2,500,000. Explanatory text (1966). FAO, Rome.

Ehrlich, W. A. 1970. The system of soil classification for Canada. Queens Printer for Canada, Ottawa.

Glinka, K. D. 1914. The great soil groups of the world and their development. Translation by C. F. Marbut. 1928. Edwards Bros., Ann Arbor, Mich.

Glinka, K. D. 1931. Treatise on soil science. Translated by A. Gourevitch. 1963. Israel Program for Scientific Translations, Jerusalem.

Hallsworth, E. G. 1968. Perspectives in soil science. Bull. Int. Soc. Soil Sci. No. 33, p. 6–13.

Hilgard, E. W. 1860. Report on the geology and agriculture of the state of Mississippi. Printed by order of the Legislature, E. Barksdale, State Printer, Jackson, Miss.

Hilgard, E. W. 1912. Soils, their formation, properties, composition and relations to climate and plant growth. MacMillan, New York.

Kubiena, W. L. 1953. The soils of Europe. Consejo Superior de Investigations Cientificas, Madrid.

Marbut, C. F. 1928a. Classification, nomenclature and mapping of soils. Soil Sci. XXV(1):51–60.

Marbut, C. F. 1928b. Soils, their genesis and classification. Published by Soil Sci. Soc. Amer. 1951.

Mill, J. Stuart. 1891. A system of logic, 8th ed. Harper & Bros., New York.

Ramann, E. 1911. Bodenkunde, Verlag Julius Springer, Berlin.

Rozov, N. N., and Ye. N. Ivanova. 1967. Classification of the soils of the USSR. Soviet Soil Sci., no. 3, p. 288–300.

Sibertzev, N. M. 1900. Selected works. Vol. 1, Soil Science. Translated by N. Kaner. 1966. Israel Program for Scientific Translations, Jerusalem.

Soil Survey Staff. 1960. Soil classification: A comprehensive system—7th approximation. Soil Conservation Service, U. S. Dep. of Agr. U. S. Government Printing Office, Washington, D. C.

Soil Survey Staff. 1967. Supplement to soil classification system—7th approximation. Soil Conservation Service, U. S. Dep. Agr. U. S. Government Printing Office, Washington, D. C.

Soil Survey Staff. 1973. Soil taxonomy, a basic system of soil classification for making and interpreting soil surveys. Soil Conservation Service, USDA. Agriculture Handbook No. 436.

Taylor, N. H., and I. J. Pohlen. 1962. Soil survey method. New Zealand Soil Bureau Bull. 25. Lower Hutt, New Zealand.

Soil Water Flow Theory and Its Application in Field Situations[1]

A. KLUTE[2]

ABSTRACT

The soil water flow theory based on the Darcy equation provides a first approximation to the description and prediction of soil water behavior. Solute-water and heat-water interactions may in some situations produce significant deviations from the Darcy-based flow theory. Methods of application of flow concepts range from their use as general, qualitative background, through various approximate uses of flow equations, to full-scale detailed prediction of the behavior of a given field flow situation. The latter is not generally feasible because of the cost and lack of the detailed knowledge of the pertinent hydraulic properties of the soil. The lack of rapid, reliable, routine methods of assessing the water flow properties of soils, and the difficulties of coping with the spatial and temporal variability of these properties are important barriers to the quantitative application of the flow theory.

INTRODUCTION

The influence of the content and energy status of water in soil upon almost every physical, chemical, and biological soil process is at least qualitatively obvious to anyone concerned with the formation and development of soils, the growth and development of plants, and the mechanical properties of soil of engineering importance. The water content and associated energy state of soil water at a given location are basically determined by (i) the physical processes of transport of water and (ii) the biological, chemical, and physical processes and reactions that produce or consume water. A large fraction of the water that falls on the surface of the soil as rain or irrigation moves into and through the soil during the processes of infiltration, drainage, evaporation, redistribution within the soil, and water uptake by plant roots. A major part of all of these phenomena involves flow of water in unsaturated soil. Because water influences so many of the reactions and processes in soil it is obviously important that we develop an understanding of its transport. The purpose of this paper is to present a brief survey of some of the concepts of soil water movement and an evaluation of the problems involved and complications encountered in applying these concepts to water flow in field flow situations.

---

[1] Contribution from the Western Region, Agricultural Research Service, U. S. Department of Agriculture and the Colorado State University Experiment Station, Fort Collins, Colorado. Scientific Journal Series 1709.

[2] Research Soil Scientist, USDA, and Professor, Department of Agronomy, Colorado State University, Fort Collins, Colorado.

SOIL WATER FLOW CONCEPTS

The theory of transport of soil water has recently been the subject of a number of reviews and books (Bear, 1970; Childs, 1967, 1969; Cary & Taylor, 1967; Klute, 1969; Miller & Klute, 1967; Philip, 1969b, 1970; Hillel, 1970; Swartzendruber, 1966). It is not necessary nor desirable to review this material in full detail. Instead a resumé of selected aspects of the theory will be given. The reader should consult the above references for a fuller exposition of the various facets of the theory.

The movement of water in a porous medium such as soil may be analyzed utilizing the continuous medium approach in which the various relevant physical properties (densities, velocities, viscosities, concentrations, etc.) are considered to be continuous functions of position and time. This is in contrast to an analysis based on a molecular viewpoint. The continuum viewpoint as used in the study of porous media may be split into two approaches (i) the microscopic, and (ii) the macroscopic. In the microscopic approach, a description of the detailed behavior of water within the pores is attempted. All physical variables involved are specified as the local "within-the-pore" values. An example of this approach is the (hypothetical) application of the Stokes-Navier equation for the viscous flow of a fluid to the problem of finding the fluid velocity pattern within the pores.

The macroscopic viewpoint utilizes physical variables that are assumed to be continuous functions of position and time in a larger sense. The fine detail of the distributions within the pores is replaced by macroscopic variables, which are volume averages of the distributions of the corresponding microscopic variables. The average must be taken over a sufficient number of pores so that the value of the variable on the macroscopic level can be considered to apply to a differential volume element located at a "point" in the medium in the macroscopic sense.

The mass flux density of water in a nonrigid porous medium, as seen by an observer in an arbitrary fixed frame of reference utilizing the macroscopic point of view, is composed of six contributions: (i) the bulk, convective transport of the solution phase relative to the solid phase, (ii) the bulk, convective transport of the gas phase relative to the solid phase, (iii) diffusion and dispersion in the solution phase, (iv) diffusion and dispersion in the gas phase, (v) convective transport of the water in the solution phase due to the motion of the solid phase, and (vi) convective transport of the water in the gas phase due to the motion of the solid phase. If the solid phase acts in a rigid manner, the last two contributions may be made zero and ignored by selecting a frame of reference attached to the solid matrix. The convective motions of the solution and gas phases are opposed by viscous forces and driven by mechanical forces. The mechanical forces include gravity, the pressure gradient force, and (in the case of the solution phase, at least) a contribution from adsorptive forces. The molecular diffusion of the water in the solution and gas phases results in a motion of the water relative to the mean motion of the phase and is quite often attributed to a driving force expressed as a concentration gradient, but may more precisely be considered as due to a gradient of thermodynamic potential. Hydrodynamic dispersion in the solution and gas phases is displayed whenever there is convective motion of the phase relative to the solid in the presence of concentration gradients in the phase. Dispersion is the macroscopically observed result of interaction between the complicated microscopic pattern of fluid velocity within the pores, the molecular diffusion produced by concentration gradients on the microscopic scale, variations in fluid properties such as density and viscosity, chemical and physical processes within the solution and gas phases, and interactions between the phases (Bear, 1969).

The mathematical-physical approach to the analysis and description of soil water flow on the macroscopic level involves (i) the selection of relations between the components of the mass flux of water and the appropriate driving forces, and (ii) the combinations of these flux equations with the mass balance equation for water. A partial differential equation is then obtained which, in principle, may be solved, subject to boundary and initial conditions, to predict the behavior of the water in a given flow situation. In the development of the flow equation numerous simplifying assumptions are made, i.e., the actual soil is represented by an idealized model. The complexity of soils forces one to make these simplifications. In building the mathematical model one tries to incorporate those features of the soil that are most important for the phenomenon being studied.

If it is assumed that

1) the soil matrix acts in a rigid manner,
2) the gas phase is at constant total pressure so that convective transport of water vapor is negligible,
3) there is negligible storage of water in the gas phase,
4) the concentration of solutes in the solution phase is negligible,
5) isothermal conditions prevail,
6) the solution phase is of constant macroscopic average density,
7) the volumetric flux of the solution phase is proportional to the hydraulic gradient (Darcy's law),
8) the gas phase diffusion flux is proportional to the gradient of the water vapor density (Fick's law),
9) the medium is isotropic with respect to Darcy flow and vapor diffusion,
10) the water vapor density in the gas phase is a defined function of the soil water pressure head,
11) the coefficient of proportionality in the Darcy equation (the hydraulic conductivity) is independent of the hydraulic gradient and position but may depend on other variables such as water content and temperature,
12) the coefficient of proportionality in Fick's law of diffusion for the water vapor (the diffusivity) is independent of the vapor density gradient and position, but may depend on other variables such as

volumetric air content, temperature, gas phase pressure, and air-filled pore space geometry,

then the following equation of flow for water may be obtained:

$$\frac{\partial \theta_L}{\partial t} = \frac{\partial}{\partial z}(K \frac{\partial h}{\partial z}) + \frac{\partial K_L}{\partial z} \qquad [1]$$

in which K_L is the hydraulic conductivity (usually assumed to be a function of the volumetric solution content θ_L), K is the combined hydraulic and "vapor" conductivity, h is the pressure head of the soil solution phase, and z is a vertical coordinate, positive upwards.

The mechanisms of transport that are assumed to be significant in arriving at equation [1] are (i) isothermal vapor transport by molecular diffusion, and (ii) convective motion of the solution phase driven by mechanical forces. Diffusion and dispersion in the solution phase, dispersion and convective motion in the gas phase, and convective transport of water vapor and water in solution due to deformation and flow of the gas phase are assumed to be zero or negligible.

Isothermal vapor diffusion, which is formally included in equation [1], has been shown by Jackson (1964a, 1964b) and Rose (1963a, 1963b) to be described by a Fick's law expression with the gradient of the water vapor density as the "driving force." The relative importance and absolute magnitude of vapor movement increases as the water content decreases. Rose concludes that, in the range of water contents important to plants, the only significant transport is liquid transport. In most applications of equation [1] little distinction is made between K and K_L, and this seems justified except in very dry soils.

If it is further assumed that there is a defined relation between the solution content θ_L and the pressure head of the soil solution h, and that this relation is not dependent upon position, then one can easily arrive at the pressure head form of the flow equation:

$$C\frac{\partial h}{\partial t} = \frac{\partial}{\partial z}(K\frac{\partial h}{\partial z}) + \frac{\partial K}{\partial z} \qquad [2]$$

or the water content form:

$$\frac{\partial \theta}{\partial t} = \frac{\partial}{\partial z}(D\frac{\partial \theta}{\partial z}) + \frac{\partial K}{\partial z} \qquad [3]$$

with the water capacity C defined as

$$C = d\theta/dh \qquad [4]$$

and the soil water diffusivity as

$$D = K(dh/d\theta) = K/C. \qquad [5]$$

The water capacity, hydraulic conductivity, and soil water diffusivity which we shall refer to collectively as the "hydraulic functions" are all to be regarded as functions of the volumetric solution content (or loosely, the water content). Since a defined relation between θ and h has been assumed, C, K, and D may also be considered as functions of the pressure head of the soil water. Equations [2] and [3] are the usual forms of the equations for the Darcy-based theory of water flow in unsaturated soils.

APPLICATION OF FLOW THEORY

Objectives and Methods

The ultimate objective of application of flow theory to field flow situations is intelligent management of the soil water regime for the benefit of man. In the process of developing the theoretical concepts that model the soil water flow system, we may increase our understanding of the system. Theoretical concepts can in principle be used in making predictions of the response of a soil water flow system to the imposition of particular boundary conditions or modifications of the properties of the system. If we can make adequate predictions we may then be in a position to modify or manage the behavior of the soil water system.

In a broad way, there would seem to be two purposes for the quantitative analysis and prediction of the performance of a given flow system (i) the verification of the validity of the flow theory, and (ii) the practical prediction of the hydraulic performance of a given body of soil material. We are concerned here with the second purpose.

Application of flow theory takes many forms ranging from qualitative use of the concepts to gain insight into the general behavior of the soil water, through various degrees of approximate, semiquantitative uses of the theory, to the more "exact" analytical and numerical solutions of the flow equation(s).

The theory of soil water movement is often used in a qualitative manner. As the basic concepts of flow in unsaturated soils become more generally known and understood by agronomist, engineers, and others dealing with the behavior of soils, it is evident that more use can and will be made of these concepts. As an example, the mere recognition that the hydraulic conductivity decreases rapidly with decreasing water content can be of great use in a qualitative way in analyzing the behavior of a soil water system.

Stating the flow theory in mathematical terms as embodied in the partial differential equations of flow provides a basis for a more quantitative prediction of behavior. The classical mathematical-physical approach requires (i) mathematical statements of the initial and boundary conditions that describe the specific flow situations, (ii) mathematical statements of the initial and boundary conditions that describe the specific flow situation, and (iii) knowledge of the conductivity and water capacity functions that characterize the soil. In principle one can then obtain a solution of the flow equation and predict the behavior of the flow system. The solution is in the form of the spatial and temporal distribution of the water content and/or the pressure head of the soil water. From these one can then derive such quantities as flux densities and cumulative flows at any point of the flow region. For example, the infiltration rate and cumulative infiltration can be determined in this manner.

The flow equation may be solved analytically or numerically. The analysis of the infiltration phenomenon developed by Philip (1969b) is an example of the analytic and semianalytic approach. For semi-infinite media with constant initial and boundary water content and with negligible effect of gravitational forces, the use of a similarity substitution (the Boltzman transformation) reduces the nonlinear partial differential diffusion equation to a nonlinear ordinary differential equation in which the water content is a function of variable $\eta = x\, t^{-1/2}$. The well-known $t^{1/2}$ and $t^{-1/2}$ dependence of various aspsects of the gravity-free infiltration follow directly from the properties of the solution of the transformed equation.

When the boundary and initial conditions are not constant, the geometry of the flow system is more complex, and/or if the medium is nonuniform with respect to the hydraulic properties, analytical solutions are not available and one must resort to numerical procedures. Numerical solutions do not lend themselves to the generalizations of results that are possible with analytical solutions. However, by the use of the concepts of similitude and the formulation of the flow equation in dimensionless variables, it is possible to effectively organize the results of numerical solutions and obtain maximum results with least effort in such an undertaking. The use of similitude concepts and dimensionless variables will greatly increase the transferability of the results of a given experiment to other flow systems of a similar type but different scale (Miller & Miller, 1956; Stallman, 1967; Corey et al., 1965). The above is also true for analytical solutions.

In recent years there have been numerous published accounts of numerical solutions of the equation of flow (e.g., see Klute, 1952; Day & Luthin, 1956; Hanks & Bowers, 1962; Rubin & Steinhardt, 1963, 1964; Whisler & Klute, 1965, 1966; Staple, 1966; Rubin, 1967; Freeze, 1969). This activity has been encouraged and made possible by the more general availability of high-speed digital, analog, and digital-analog (hybrid) computors. Many of these solutions have not been associated with experimental observations. Those solutions that have been reported in conjunction with experimental work have usually been laboratory studies, designed to explore the validity of the flow theory. These laboratory studies, although performed under idealized conditions and on ideal media in flow systems of relatively simple geometry, are not without value for application to the field. A greater understanding of the physics of the flow can be obtained from a study of flow systems under controlled conditions, and most of the basic concepts of unsaturated flow have been formulated from observations on idealized flow systems in the laboratory. The information obtained in such studies provides a useful guide in designing and interpreting field water flow experiments. The behavior of these idealized flow systems provides a point of departure for treating more complex systems.

The approximate methods occupy an intermediate position between qualitative application of flow theory and the "exact" analytical or numerical solutions. In these, certain recognized aspects of the flow are ignored in order to obtain a more tractable flow equation which can then be solved. In actual fact any analysis is approximate in one or another sense. It is a question of the degree of approximation involved. The basis on which aspects of the flow are neglected is usually one of experience and is based on observation of the behavior of the flow system. By the judicious use of approximations it is possible to arrive at results of practical utility while still retaining physically significant features of the flow. As examples of this type of approach we may cite the use of a weighted mean diffusivity to obtain the inflow or outflow behavior of a soil column or profile (Gardner, 1969b, 1962; Doering, 1965), the assumption that the water content in a drying soil column depends largely on time and not so much on position (Gardner & Hillel, 1962), the assumption of a constant suction at the wetting front during infiltration (Green & Ampt, 1911), and the assumption that water will drain from a wetted uniform soil profile under unit hydraulic gradient (Black et al., 1969; Davidson, 1969).

Gardner (1962) used the assumption of a constant mean diffusivity in deriving an equation for the volumetric outflow rate of a vertical drainage column. Youngs (1960) based his analysis of the same system on the assumption of a constant suction at the drainage front, which is the (assumed) line of demarcation between the drained and undrained soil. Jackson and Whisler (1970) modified Youngs' approach, and assumed among other things that hydraulic conductivity is linearly related to the average water content. They assert that their theory agrees with experiments over a longer time than either Gardner's or Youngs' theory,

None of these procedures is capable of predicting the water content distribution within the draining column but often the drainage rate is the primary feature of interest. In the absence of complete conductivity and water retention data for the soil material, and where computer solutions are not feasible, such analyses may be most useful. However, the drainage-time equations derived by these methods contain various parameters, some of which may not easily be evaluated a priori, and if the parameters are not available the predictive value of the equations is severely limited.

Complications

A review of the long list of assumptions involved in the development of equations [2] and [3] will convince the reader that the Darcy-based theory of flow is strictly applicable only to a very idealized medium. Field soils do not conform to the assumptions listed. Soils are anisotropic, nonuniform, nonisothermal, and sometimes water repellent. They contain entrapped air as isolated bubbles, and the gas phase may, in some circumstances, not be at constant pressure let alone atmospheric pressure. Soils may possess varying degrees of structure. Fissures, cracks, aggregates, and worm holes are common features of soil profiles. Deformation may occur as swelling with or without bulk volume change and as translocation of the finer particles suspended in the soil solution. Biological activity within the soil (plant roots and/or microbial) may produce or consume water and may also modify the pertinent hydraulic properties such as conductivity and the water capacity. Various kinds of interactions may take place between heat flow, solute flow, and water flow. The hydraulic properties such as conductivity and water capacity may and do exhibit the phenomenon known as hysteresis, and non-Darcy flow may occur.

NON-DARCY FLOW

The validity of the extended Darcy equation, with a conductivity that is a function of the water content of the medium, is a central assumption of the flow theory as expressed in equations [1], [2], or [3]. Swartzendruber (1966) has recently devoted considerable attention to the deviations from the Darcy equation that occur at low flow velocities. The deviation from linearity between hydraulic gradient and the Darcy velocity at high gradients due to the increasing importance of inertial forces is relatively well understood and is not usually a matter of concern in soils at the usual flow velocities and gradients encountered. Deviations from proportional behavior at low flow velocities are more difficult to identify, define, and associate with a causative mechanism. Experimental error, water properties, in the pores that are modified with respect to those of bulk water, streaming potential effects, osmotic effects, and particle translocation are some of the mechanisms that have been advanced as explanations for nonproportional behavior. Nonproportional behavior does occur and in some media it is very significant. As a first approximation, it seems reasonable to assume that the Darcy equation is valid, but one should be alert to the possibility that difficulties in obtaining agreement between the predictions of a Darcy-based theory and experimental observations may in part be due to the fact that the Darcy equation is invalid for the particular medium being studied. The reader should consult Swartzendruber's review for a more extended discussion and an access to the literature on this problem.

STATE OF FLOW-DEPENDENT WATER CONTENT-PRESSURE HEAD RELATIONSHIP

In most applications of equations [2] and [3] it has been tactily assumed that the water content-pressure head relation was not a function of

$\partial\theta/\partial t$, i.e., the state of flow. Water retention data obtained by conventional static hydraulic equilibrium techniques in pressure cells, and tension apparatus has been used to evaluate the water capacity function. These data have then been used to analyze and predict the behavior of unsteady-state flow systems.

In some media, at least, it appears that the static equilibrium water content-pressure head relation may not be the same as a relationship derived from unsteady-state measurements (Topp et al., 1967; Davidson et al., 1966).[3] Explanations of the dependence of the $\theta(h)$ function on the state of flow and on the size of the pressure head steps used to arrive at a given pressure head include surface tension dependence on "age" of the air-water interfaces, heterogeneities in the packing of the porous medium, and experimental errors and artifacts. The extent and reality of results of this kind are not clear at present, but one should be aware that this complication may occur.

NONISOTHERMAL FLOW

The upper part of a soil profile is easily recognized to be a nonisothermal medium. Temperature fluctuations penetrate to a depth of 20 to 30 cm and seasonal fluctuations to much greater depths (Carson & Moses, 1963). Temperature gradients of the order of 0.5C/cm are common.

The Darcy-based flow theory does not consider flow due to forces that arise because of temperature gradients. Temperature gradients in a soil produce associated vapor pressure gradients, spatial variations of surface tension, and adsorptive forces which may cause transfer of water that can be identified with the temperature gradient. In any given application of flow theory, the significance of the water transport due to nonisothermal conditions must be assessed. Consideration must be given to the magnitude and duration of application of the temperature gradient, and the magnitude of the transfer coefficient(s) associated with the temperature gradient.

In general, the temperature of the soil effects water transport through its effect on the forces that cause the water to move, and through its effect on the conductivities and diffusivities in the various flux equations.

Water flow in response to a temperature gradient is usually in the direction of decreasing temperature. In soils with a gas phase, a temperature gradient produces an associated vapor pressure gradient and a surface tension gradient. There is then a component of the vapor flux that can be related to the applied temperature gradient, and a component of the liquid flux which is associated with the temperature gradient. Less well understood is the effect of temperature on the various solid-liquid adsorptive forces that contribute to the energy status of the solution and the water in it. The probable and possible variation of the intensity of the adsorptive forces with temperature could lead to a transport of water.

The theory for thermally driven water flow in soils has been developed

---

[3]R. H. Sedgley. 1967. Water content-pressure head relationships of a porous medium. Ph.D. Thesis, Agronomy Department, University of Illinois, Urbana, Ill.

from a mechanistic point of view by Philip and de Vries (1957), and by the more abstract methods of irreversible thermodyanmics (Cary & Taylor, 1962a, 1962b; Taylor & Cary, 1964; Groenevelt & Bolt, 1969). The theory developed by Philip and de Vries considers the effect of temperature on the vapor pressure and the surface tension and, hence, on the driving forces for vapor diffusion and liquid phase flow. In their theory there is no provision for thermally driven flow in saturated media, a phenomenon which can occur (Corey & Kemper, 1961; Taylor & Cary, 1960). In the Philip-de Vries theory the equation for the volumetric flux density of water is written:

$$\bar{Q}_w = -D_T \, \nabla T - D_\theta \, \nabla \theta. \tag{6}$$

The thermal diffusivity D_T is the sum of a thermal liquid diffusivity and a thermal vapor diffusivity. The isothermal water diffusivity D_θ is the sum of a liquid phase diffusivity and an isothermal vapor diffusivity. Philip and de Vries indicate how these diffusivities may be calculated from the conductivity-water content and the pressure head-water content relations, the fluid properties, and a number of other parameters. Hysteresis and the effect of solutes on the soil water vapor pressure are ignored.

The analyses based on the methods of irreversible thermodynamics lead to a flux equation for water in a "salt free" nonisothermal soil that is qualitatively of the same form as that written by Philip and de Vries. However, the approach gives little or no insight into the method of calculation of the transport coefficients from basic fluid and soil properties.

The coefficients could as well be regarded as empirical coefficients whose values and functional dependence are to be determined experimentally. With this approach, one is not tied to a particular theory for "explaining" the coefficients in terms of other fluid and soil properties. However, it must also be recognized that even if we knew that a flux equation of the form of equation [6] was valid, but we did not have a reasonable theory for the magnitudes and behavior of the coefficients, we would not be as confident in the use of the equation and we would not always be aware of its limitations.

It can be seen that at least one more transport coefficient, viz., D_T is needed to treat nonisothermal flow as compared to isothermal flow. It is difficult enough to establish the behavior of D_θ, and the additional requirement of knowledge of the D_T function (of water content, and possibly temperature) greatly compounds the difficulties of application.

In addition to the effects of temperature upon the driving forces for flow, there are also direct thermal effects on the transport coefficients, viz., the conductivities and diffusivities. Since the hydraulic conductivity is inversely proportional to the viscosity, an increase in temperature will increase the conductivity. In saturated media, corrections for the temperature dependence of viscosity are often made, and amount to approximately a 20% increase in conductivity for an increase in temperature from 20C to 30C. For flow in unsaturated soils, Gardner (1959a) concluded that the temperature dependence of viscosity accounted for the temperature dependence of soil water diffusivity at higher water contents, but that the effects of temperature on the pressure head became more important at lower water contents. Jackson (1963) found similar results. A temperature increase from 5 to 45C approximately doubled the soil water diffusivity. Jensen, Haridasan, & Rahi (1970) have also found that near saturation the changes in conductivity due to temperature were explained by the temperature dependence of viscosity. At lower pressure heads (higher suctions), the temperature effect on water retention at a given pressure head was the major influence on the conductivity-pressure head relation.

At a given pressure head the retention of water by most porous media decreases as the temperature increases (Wilkinson & Klute, 1962; Peck, 1960). The decrease in water content is due to the decrease of surface tension with increasing temperature, to the effects of temperature on entrapped air, and possibly to the effects of temperature on the adsorptive forces.

C. W. Rose (1968a, 1968b) conducted a theoretical and experimental study of the thermal and hydraulic components of flow in the upper 15 cm of a bare soil under high radiation conditions. The theory of Philip and de Vries, with some modification, was used. The thermal vapor flux was found to be of comparable magnitude to the liquid flux and was oscillatory in direction and magnitude, being upward during the night and downward during the day. Hysteresis was neglected.

During the winter months in the northern latitudes, there is a significant upward movement and accumulation of water associated with a long-term application of relatively low temperatures at the soil surface (Willis et al., 1964; Ferguson, Brown, & Dickey, 1964). Estimates of the thermally driven flow in unsaturated soil made by Cary (1966) do not seem to yield fluxes of sufficient magnitude to explain the flow. The phenomenon of flow in frozen soils has been studied by Dirksen and Miller (1966), Hoekstra (1966), Williams (1968), and Anderson and Hoekstra (1965). It appears that water flow in frozen soils takes place under temperature gradients in the films of unfrozen water. The concepts of flow in and to frozen soils have been greatly developed in recent years. Williams (1968) gives a comprehensive review of much of the literature and the concepts that have developed.

The probable significance of thermally driven water flow has been evaluated by Cary and Taylor (1967) and Philip (1957b). Cary and Taylor conclude that normal thermal gradients in the root zone can be about 10 times as effective as gravity in moving water when the soil is at or near saturation, and will become of increasing importance as the soil dries, perhaps up to 1000 times more effective than gravity by the time the permanent wilting point is reached. Philip, on the other hand, concludes that thermally driven flow in the evaporative drying process is of minor importance until the soil becomes very dry. It seems that if one is concerned with flow in the upper 20 to 30 cm of the soil profile, one should be aware of the probable contribution of thermally driven flow to the total flow. On the other hand, if the problem being studied involves water flow at depths greater than about 20 cm and over periods of time less than a few weeks, one can probably safely ignore the thermal effects. Thermal effects cannot be ignored in long-term flow studies as shown by the observations of Willis et al. (1964) and Ferguson et al. (1964). Quantitative data on transfer coefficients for thermal water flow are relatively scarce and assessment of the importance of its contribution to the total movement of water in soils continues to be a subject of research. The measurement of the water flux components in the presence of diurnal temperature fluctuations should be continued. For the present, it seems that the application of the available concepts of nonisothermal flow to field situations will be of a qualitative nature, in view of the problems of determining the necessary transport coefficients and in solving the equations of transport for thermal movement.

SOLUTE-WATER-SOIL MATRIX INTERACTIONS

The soil solution is never homogeneous. Extraction of nutrients by plants, evaporation of water at the soil surface, and the application of fertilizers are but a few of the reasons that solute concentration gradients exist in the soil solution.

The effect of the solute concentration and composition, and gradients thereof, upon water flow in soils may be broadly classified into (i) effects of the driving forces for flow and (ii) effects on the transfer coefficients.

Whenever a gradient of solute concentration is present, there is also a gradient of the chemical potential of the water in the soil solution. If a gas phase is present there will be a contribution to the vapor pressure gradient and hence an effect on the vapor flux. In the liquid phase, molecular diffusion will occur. If the latter occurs in the presence of bulk flow of the solution phase, hydrodynamic dispersion will also occur. Whenever the soil exhibits a degree of selectivity in its action on the solute versus its effects on the water, the phenomenon of salt sieving or reverse osmosis will be displayed. In qualitative terms when the solute is not as free to move into or through the medium as is the water, an osmotic type of flow can occur. One reason for the selective action of soils upon solutes is the anion exclusion from the fine pores and thin films of solution in clay systems. The soil exhibits a degree of "membrane" action. The ideal selective medium is the classical semipermeable membrane. Osmotic flow in and through clay plugs has been the subject of a considerable amount of study (Letey, Kemper, & Noonan, 1969; Kemper & Rollins, 1966; Abd-El-Aziz & Taylor, 1965). Most of this work has been done with two-phase clay systems and relatively little in three-phase systems.

The flux equation that has been applied to osmotic flow across clay membranes or plugs is of the form:

$$\bar{Q} = -K(\nabla H - \sigma \nabla \pi) \qquad [7]$$

where the osmotic efficiency coefficient σ has been introduced as a multiplier of the osmotic pressure gradient $\nabla\pi$, and ∇H is the hydraulic gradient. The osmotic efficiency coefficient ranges from zero to unity. In a perfect semi-permeable membrane σ has the limiting value of unit. A medium with no selective action toward the solute has an osmotic efficiency coefficient of zero.

The significance of salt sieving, reverse osmosis, or osmotic flow in soils is not entirely clear. In laboratory studies of flow across clay plugs the phenomenon is certainly significant and observable. But in the field, one is more often concerned with flow within the porous medium. In nonuniform media such as those cases where fine-textured layers are interspersed with coarser media, the phenomenon of salt sieving must be considered if the fine-textured layers have an osmotic efficiency coefficient significantly different from zero. Nielsen et al. (1972) recently reviewed the literature on salt sieving.

The effect of the solute concentration and ionic composition on the hydraulic properties of soils and clays has also been the subject of considerable research (Reeve, 1960; McNeal & Coleman, 1966; Rhoades & Ingvalson, 1969; Yaron & Thomas, 1968). The most important factors that affect the conductivity and diffusivity seem to be the electrolyte concentration and the exchangeable sodium percentage. The effect of these two factors is rather similar for both saturated and unsaturated flow (Gardner et al., 1959; Quirk & Schofield, 1955). When a solution that differs from the soil solution is passed through a soil, large changes in hydraulic conductivity, perhaps as much as by a factor of 100 to 1000, can occur. These changes occur because of the rearrangement of the solid matrix brought about by the change in its chemical environment upon injection of the displacing solution. More often than not the water that is applied as rain or irrigation water is not at chemical equilibrium with the soil solution. Christansen and Ferguson (1966) have studied infiltration using water of different composition than the initial soil solution and found observable deviations from $t^{1/2}$ behavior.

Several researchers have evaluated the significance of solute-induced flow (osmotic, diffusive, and dispersive). Letey (1968) and Letey et al. (1969) observed that in most soils the osmotic flow that occurs at low suction is small compared to the usual nonosmotic, hydraulic, or Darcian flow. At higher suctions, the hydraulic flow decreases in magnitude and the osmotic flow, while still small, becomes relatively more important. Cary and Taylor (1967) and Kemper (1961) state that in normal field soil conditions, the solute-induced flow is not significant until the suction is 1 bar or higher. It appears that within the soil, in the range of water content of significance to plants, solute concentration gradients will not be a significant factor in water movement. Exceptions to this would be (i) at evaporation sites such as the soil surface, (ii) at freezing surfaces (ice lenses), and (iii) at sites of local application of fertilizer salts. The significance of solute-induced flow and its mechanisms continues to be a subject for research, especially the unsteady-state aspects of such flow and in unsaturated media.

FLOW IN DEFORMABLE SOILS

Most soils do not behave as rigid media, i.e., they usually undergo at least some deformation as water flow occurs in them. The drag of the flowing fluid may induce a rearrangement of the finer particles. As the suction is increased, the soil particles, especially the clays, are drawn together with a resultant decrease in void ratio and with important consequences for the hydraulic conductivity. These changes in particle arrangement may be reversible or irreversible.

The problem of consolidation of clays has been a matter of some concern in soil mechanics (e.g., see Terzaghi, 1943; Biot, 1956). Soil physicists have tended to avoid the problem of flow in deformable media. Recently, however, papers have begun to appear in the literature on various aspects of such flow (e.g., Philip, 1969a; Philip & Smiles, 1969).

Philip and Smiles (1969) have analyzed flow in deformable media such as soil of high colloid content. The hydraulic conductivity, the pressure head, and the void ratio were assumed to be known functions of water content. The Darcy law was assumed to be valid, with the recognition that it applies to solution phase flow relative to the solid matrix. The combination of the Darcy equation with the continuity requirement in a material coordinate system leads to a flow equation of a nonlinear heat conduction form. In developing the equation, it was assumed that water flow due to solute concentration gradients was negligible and that salt sieving was of no consequence. Hysteresis and gravity were also neglected.

Because of the mathematical form of the flow equation, the following factors that are predicted for flow into a rigid semi-infinite medium at constant initial and boundary water contents are also predicted for such flow in the swelling medium: (i) square root of time dependence of infiltration rate, (ii) cumulative infiltration, and (iii) position of a given water content.

In the analysis it was assumed that no salt-sieving effects occur at the interface between the porous medium and the body of solution which is absorbed or desorbed by the medium, and that the solution that is absorbed or desorbed is the same as the solution in the pores of the medium. If the composition of the solution entering the medium differs from that already in the medium, the hydraulic properties of the medium will change and the behavior will deviate from the square root of time dependence (e.g., see Christensen & Ferguson, 1966).

ENTRAPPED AIR

When a soil is wetted rapidly to zero suction, complete saturation is seldom, if ever, attained immediately. Up to 25% of the pore space may contain a gas phase in the form of isolated bubbles and irregular pockets. The geometry of this entrapped air is complex and unknown in detail. Due to the curvature and surface tension of the air-water interface, the pressure in the entrapped air exceeds that in the solution phase. If the pressure in the bubbles exceeds that of the external atmosphere (as will be the case when a soil is wetted by infiltration), the bubbles are not stable, and the gases they contain will dissolve and diffuse or be convected to the exterior of the sample. In this way, complete saturation will be attained eventually, but the process is very slow and for those flow situations having a relatively short time scale the solution of the bubbles can be neglected.

The presence of entrapped air introduces several complications to the construction of a valid flow theory. The entrapped air plays a role in the $K(\theta)$ and the $\theta(h)$ functions. Fluctuations of pressure in the solution phase and of temperature of the medium will cause changes in the pressure in the entrapped air, and these will have an effect on the $\theta(h)$ and $K(\theta)$ functions. The gradual solution of the entrapped air will also cause these functions to appear to be time dependent.

The mathematical description of the behavior of the entrapped air and its effects on the hydraulic properties is not in a very satisfactory state, largely because of the complex geometry involved. In most applications to field situations of flow, the best that can be done is to utilize $K(\theta)$ and $\theta(h)$ functions that are determined under wetting conditions similar to those that will be found in the field situation, and to ignore the time dependence of these functions due to the solution of the entrapped gas. For rapidly varying flow systems this will be a reasonable approach, but in slowly varying flow systems difficulties may be expected in such an approach.

The phenomenon of entrapped air has been studied by Peck (1965), Bloomsburg and Corey (1964), and DeBacker (1967). Its significance in various flow situations is still a matter for research.

GAS-PHASE FLOW

In the development of equations [2] and [3] the viscous flow of the gas phase was ignored. Infiltration of water into a rigid porous medium requires the displacement and flow of the gas phase. Due to the low viscosity of the gas phase and to the fact that in many situations the displaced gas has free access to the exterior of the medium, the gas phase pressure is very nearly constant and equal to the ambient pressure. However, there are situations where the gas phase cannot freely escape. Infiltration of water due to capillary and gravitational forces will then cause an increase in gas phase pressure which retards infiltration. The increased gas phase pressure may cause the gas to displace water from the larger pores of the medium and escape to the exterior of the sample. Associated with this escape there may also be a disruptive action on the solid matrix, e.g., a separation of the upper part of a flow column from the rest of the flow system.

The gas phase flow is due to a total gas phase pressure gradient. Unless the conductivity of the medium to the gas phase is relatively low, the pressure in the gas phase can often be assumed as spatially constant, but variable with time due to the compression by the infiltrating water. Gas phase pressures up to 100 cm of water in excess of atmospheric pressure have been observed (Wilson & Luthin, 1963; Garner, Donaldson, & Taylor, 1969).

The effect of back pressure in the gas phase has been studied by Wilson and Luthin (1963) and by Peck (1965). Equations for the simultaneous flow of the gas and solution phases have been formulated[4] (Elrick, 1961; Green et al., 1970) and some progress has been made in solving them, at least by numerical techniques.

For large-area ponded infiltration into a soil profile with a water table or impermeable layer at a shallow depth, the back pressure developed in the gas phase cannot be ignored. In cases where there is a free lateral flow path for the gas, or where there is no barrier to gas flow near the surface, it is reasonable to ignore the gas phase flow and assume that it is at atmospheric pressure. In a field situation where predictions of the water flow behavior may be required, it becomes necessary to characterize the mode of escape of the gas phase in order to predict the gas phase pressure buildup. Satisfactory modeling of such features in those instances where the resistances and pathways for gas escape are unknown appears difficult if not impossible.

HYSTERESIS

The cyclic nature of the boundary conditions at the surface of a soil profile, with periods of water application by rainfall or irrigation alternating with evaporative periods, induces fluctuations of water content and soil water potential in the upper part of the profile. In any flow system in which the time sequence of water content (or pressure head) undergoes a reversal, i.e. $\partial\theta/\partial t$ changes sign, the phenomenon of hysteresis is involved. Hysteresis is the term applied to designate the fact that the water content depends not only on the pressure head, but on the past history or the sequence of pressure heads to which the soil water has been subjected. Wetting $\theta(h)$ curves display a lower water content at a given h than drying curves, and there is also an infinity of scanning curves, both wetting and drying. Hysteresis, real or apparent, has a number of possible causes including (i) contact angle dependence on whether the medium is wetting or drying, (ii) the cellular nature of pore space (in some media) and the associated possibility of multiple stable positions of an air–water interface in such pores, (iii) the effects of air entrapment, and (iv) particle rearrangement. In some cases, it is possible that observed hysteresis is due to a lack of equilibrium, but this cannot be invoked as a general explanation.

The hysteretic behavior of the $\theta(h)$ relation makes it difficult to determine the water content at a given pressure head and vice versa. The hysteretic $\theta(h)$ relation introduces considerable difficulty in solving the flow equation, and analytical and semianalytical means to cope with the hysteretic behavior are not available at present.

---

[4]LeVan Phuc. 1969. General one-dimensional model for infiltration. M.S. Thesis, Department of Civil Engineering, Colorado State University, Fort Collins, Colo.

With the increased availability of high-speed computers, problems involving one or another aspect of hysteresis have been attacked using numerical techniques. Whisler and Klute (1965) analyzed the case of infiltration into a vertical column assumed to be initially at equilibrium under gravity. Each point of the column then rewetted along a wetting scanning curve that departed from the main drying curve at a pressure head equal to the negative of the elevation of the point. In a sense, the problem was one dealing with a nonuniform medium, with the nonuniformity induced by the hysteretic nature of the $\theta(h)$ relation. However, there were no reversals of the algebraic sign of $\partial\theta/\partial t$ in the system.

Rubin (1967), Staple (1969), Hanks, Klute, and Bresler (1969), and Phuc[4] have developed computer solutions for infiltration followed by redistribution. In these solutions it was assumed that the hysteretic nature of the $\theta(h)$ function was known in sufficient detail so that by interpolation, using a given pressure head and knowledge of the time sequence of pressure heads at a given point in the flow system, the water content could be determined.

Theories of the nature of hysteresis have been utilized in an attempt to calculate the scanning curve relations from a limited amount of information about the medium. The independent domain theory has been studied by Poulovasillis (1962), Topp and Miller (1966), and Topp (1969) as well as others. The results have been somewhat contradictory. Poulovasillis finds the theory to be applicable; Topp and Miller and Topp do not. The theory, if valid, would provide a means of calculation of the primary drying scanning curves from the primary wetting scanning curves, or vice versa, and would provide for a considerable reduction in labor for measurement of the hysteretic $\theta(h)$ function. Philip (1964) has described a similarity hypothesis which seems at least in a general way to be equivalent to the independent domain theory. Ibrahim and Brutsaert (1968) have used the independent domain theory in conjunction with their numerical solution for hysteretic flow. Topp (1971) has described a modification of the independent domain theory which includes certain interaction features between various pore size classes. The fit to experimental data seems better, but the theory is more complex to use, and more input data are needed to use it. It seems that the theory of hysteresis has not yet been brought to the stage where it can effectively be used in analyzing and predicting the behavior of flow systems.

It is evident from the available data on the water retention of various soil materials that the $\theta(h)$ function is hysteretic, and especially so at pressure heads near zero (low suction). The mapping of the hysteresis in the $\theta(h)$ relation is a formidable task, and especially so in field soils. It is possible that for specialized purposes in laboratory flow studies, the scanning curve information can and would be determined, but for field application purposes such measurements by presently known techniques do not appear feasible.

The evidence for the hysteretic nature of the conductivity function is less clear. In some media, $K(\theta)$ was found to be hysteretic (Poulovasillis, 1969) but in other media little or no hysteresis in $K(\theta)$ was found (Elrick & Bowman, 1964; Topp & Miller, 1966). Even if $K(\theta)$ is nonhysteretic, the function $K(h)$ will be hysteretic because of the hysteretic nature of $\theta(h)$. It would seem to be reasonable to assume that $K(\theta)$ is nonhysteretic for most practical purposes, unless direct evidence to the contrary is available.

It is desirable to establish the extent of the deviations and perturbations caused by hysteresis and to search for and develop simplified means of coping with hysteresis in the analysis of flow systems. Evaluations (on the basis of the intended application of flow theory) must be made of the errors introduced either by neglect of hysteresis or by the use of an approximate treatment of it. In some cases of flow, one can be quite sure, from knowledge of the boundary and initial conditions, that hysteresis either cannot be involved or if it is, the extent to which it would be displayed would be minimal. In field soils, with nonuniform initial conditions and cyclic boundary conditions, hystersis must play a role in the behavior of the flow. The question to be answered clearly is "how much?"

NONUNIFORM MEDIA

In most applications of flow theory the hydraulic properties of the medium, $K(\theta)$ and $\psi(\theta)$, have been assumed to be the same at all positions in the flow system. In a soil profile, it is rarely, if ever, that the entire profile can be characterized by a single conductivity function or a single water retention curve. The flow equation can be written for one-dimensional vertical flow as:

$$C_H(h, z)\frac{\partial h}{\partial t} = \frac{\partial}{\partial z}\left[K_H(h, z)\frac{\partial H}{\partial z}\right] \qquad [8]$$

where the water capacity and hydraulic conductivity are interpreted as hysteretic functions of pressure head and position, $C_H(h, z)$ and $K_H(h, z)$. Use of the flow equation in the above form assumes that the spatial dependence of the hydraulic functions is known. While this is possible in principle, in practice it may be difficult to establish the spatial dependence. Certainly the problem of measurement of the hydraulic properties is increased by the requirement that their spatial dependence be specified.

Hanks and Bowers (1962) and Whisler and Klute (1966) have used numerical procedures to develop solutions for infiltration into layered soils. The qualitative features of the infiltration behavior that have been observed experimentally (e.g., see Miller & Gardner, 1962) were predicted by the numerical solutions. In particular, Whisler and Klute (1966) showed that the numerical solution gave a qualitative prediction of the experimentally observed "hold up" on the advance of the wetting front as it goes from a fine- to a coarse-textured layer in an infiltration.

Philip (1967) has discussed the problem of sorption and filtration in heterogeneous media and has developed a quasi-analytical treatment of "scale-heterogeneous" media, i.e. those media in which the characteristic microscopic internal length scale varies spatially.

STRUCTURED MEDIA

In the development of the equations of flow, it is assumed that the conductivity and other hydraulic variables are defined on the macroscopic level and that there is local equilibrium on what may be called the Darcy scale. In unsteady flow in aggregated media, which may be regarded as composed of macroporous and microporous regions, there will be a disequilibrium between the water in the aggregates (microporous) and between the aggregates (macroporous). The pore size distribution in such media is bimodal, i.e. there tends to be two major size classes of pores that are dominant. Philip (1968a, 1968b) has analyzed one-dimensional desorption into a semi-infinite aggregated medium. From his analysis he concludes that the medium behaves initially as a classical uniform medium with a sorptivity S determined by the macroporosity, and that the sorptivity increases with time, and eventually approaches a limiting value greater than S. Philip concluded that the characteristic time for absorption into the microporosity was small and that deviations from diffusion analysis would occur only during a short initial period.

Another kind of commonly occurring structure in porous media is that of cracks and fissures such as those created during the drying of swelling soils. Flow into or out of these media may be treated on two levels. In the first, the geometry of the cracks or fissures and boundary conditions at their surfaces are specified and flow into or out of the soil is analyzed by a multidimensional application of the Darcy-based flow theory. In the second, no attempt is made to specify detailed boundary conditions at the surfaces of the fissures, but instead a large scale macroscopic viewpoint is adopted in which, in a sense, the details of the geometry of the cracks is a microscopic consideration. The conductivity function, water capacity function, and the void ratio-water content function which are necessary to characterize such soils (Philip & Smiles, 1969) must be based on the whole soil and not just on the soil between the cracks. Relatively little seems to have been done on the problem of unsaturated flow in fissures or cracked soils.

WATER REPELLENT SOILS

Hydrophobic substances can drastically change the hydraulic properties of soils, primarily through modification of contact angle and surface tension. There has been extensive research on the effects and nature of hydrophobic materials in soils (e.g., see Debano & Letey, 1969; Watson & Letey, 1970). Debano (1971) has investigated the effects of hydrophobic substances on horizontal and vertical infiltration. Diffusivities for water-repellent soil were smaller at all relative water contents than for wettable soils. The infiltration rates for the water-repellent soils were lower. The diffusivity analysis indicated that the hydrophobic substances had the greatest effect at lower water contents. Debano concluded that equation [3] was not very successful as a model for vertical infiltration in either wettable or nonwettable soils.

It is likely that the manner in which these substances are distributed in the soil, laterally and vertically, will add to the nonuniformity of the soil with respect to its hydraulic properties. Whether or not flow theory can be successfully applied to such materials is an open question.

Applications to Field Flow Systems

In a few cases solutions of the flow equation have been compared with field observations. Green, Hanks, and Larson (1964) used the numerical procedure of Hanks and Bowers (1962) to solve the flow equation for a layered soil profile. The necessary soil hydraulic properties were obtained by laboratory methods. The calculated infiltration rates were in fair agreement with the experimentally observed rates. Wang and Lakshiminarayana (1968) calculated water content changes for infiltration and drainage using conductivity data for each identifiable soil layer of Panoche clay loam as measured in the field by Nielsen et al. (1964). Excellent agreement between calculated and field-measured distributions of water content, cumulative infiltration, and soil water flux was obtained.

Green et al. (1970) obtained numerical solutions of equations for the flow of water and air for infiltration and subsequent evaporation. Soil hydraulic properties were estimated from the various results available in the literature for soils similar to those in the field test site. The predicted soil water profiles were in good agreement with the measured profiles. Nielsen, Kirkham, and Van Wijk (1961) used the method of Philip (1955, 1957a) to calculate moisture profiles for infiltration into two silt loam soils using diffusivity data measured by laboratory methods. The agreement between theoretical and experimental curves was fairly good in one case and rather poor in another.

Approximate methods of analyses have been used to predict soil water behavior in field soils. For example, Black et al. (1969) used the assumption that the hydraulic gradient during drainage from a uniform soil is approximately unity and that water will drain almost at the same rate from all depths to derive an equation for the water flux at a given depth. The equation was applied to a Plainfield sand in a lysimeter with good success in predicting changes in soil water storage. Davidson et al. (1969) applied the same kind of analysis to several soils varying in uniformity as displayed by the soil water content and pressure head profiles during drainage. The comparison between the experimental and theoretical drainage rates was very good for the more uniform soils, but was less so for a nonuniform soil profile. The authors indicated that for nonuniform soil profiles and for those cases where there is a large change in water content near the soil surface due to evaporation, computer solutions similar to those of Wang and Lakshiminarayana (1968) are needed to predict the drainage and evaporation rates. Black, Gardner, and Tanner (1970) used an average hydraulic conductivity in a hydrologic balance equation with the Darcy equation to calculate the drainage from the root zone of snap beans (*Phaseolus vulgaris* L.) grown on a Plainfield sand. The results compared favorably with measured drainage rates in a lysimeter.

Quantitative application of the Darcy-based flow theory requires knowledge of the hydraulic conductivity and water retention functions for the flow medium. The literature in soil physics is plentifully supplied with papers on methods of measuring the diffusivity, conductivity, and water retention functions (Doering, 1965; Bruce & Klute, 1956; Gardner, 1956; Miller & Elrick, 1958; Richards & Moore, 1952; Tanner & Elrick, 1958; Klute, 1965). These methods are inherently laboratory methods.

Field estimates of the hydraulic conductivity function have been made by Richards, Gardner, and Ogata (1956) and Ogata and Richards (1957) using tensiometer and gravimetric soil water content data. Brust, Van Bavel, and Stirk (1968) and Van Bavel, Stirk, and Brust (1968) used tensiometry and the neutron method to measure hydraulic head and water content distributions in a field soil profile. Hydraulic conductivities were obtained from hydraulic gradients obtained from the tensiometer measurements, and water flux was calculated from the water content profiles. Nielsen et al. (1964) and La Rue, Nielsen, and Hagan (1968) used tensiometers to measure hydraulic gradients and obtained water content data from soil water retention data measured in the laboratory. Rose, Stern, and Drummond (1965) used water content data obtained by the neutron method and water retention data by laboratory methods to find the hydraulic gradient and water flux, and hence calculate the hydraulic conductivity-water content function in a loam soil. It seems a priori better to measure the hydraulic gradient rather than infer it from a water retention curve. This is supported by the studies of Rose et al. (1965), Nielsen et al. (1964), and LaRue et al. (1968).

The choice at the present time seems to be whether it is better to take a few samples to the laboratory and make measurements on them under more controlled conditions, or to make measurements *in situ* under less controlled conditions. The laboratory measurements may give more accurate and precise results on the samples as they exist in the laboratory than the field methods. However, the field methods are likely to give results that are more representative of the soil as it exists in the field. Measurements in the laboratory on "undisturbed" core samples suffer from the adverse sampling situation imposed by the small size and number of the core samples that can be processed, and the possible and probable changes in soil properties that can occur when a sample is removed from the field to the laboratory. For such reasons, *in situ* measurement of the properties would seem to offer promise as a means of hydraulic characterization of a field site.

Because of the difficulties attendent to the measurement of the $K(\theta)$ [or $K(h)$] function there has been considerable interest in estimating the conductivity from some other property of the medium which, it is hoped, would be easier to measure. Several procedures have been developed to calculate the conductivity from the water content-suction relationship. These have been studied experimentally by Jackson, Reginato, and Van Bavel (1965); Green and Corey (1971); and by Nielsen, Kirkham, and Perrier (1960). Jackson et al. (1965) concluded that the method of Millington and Quirk (1959, 1961) combined with a matching factor obtained from one measured conductivity value (e.g., at saturation) gave satisfactory results. Green and Corey (1971) developed a revision of Marshall's (1958) procedure and compared it and the original procedure of Marshall and that of Millington and Quirk. It was concluded that any of the methods would give satisfactory results if they were combined with one measured conductivity value.

CONCLUDING REMARKS

An evaluation of the validity of any given theory will be highly dependent upon the intended use and application, and also upon the degree of rigor required by the evaluator. Evaluation is a continuing process—it will not be completed by performing one or a few experiments by one or a few individuals. The accumulation of evidence both pro and con will continue, and as it does, gradually a clearer understanding of the soil water flow system will emerge and the limitations and degree of validity of such a theory as the Darcy-based flow theory will be better understood.

Some experiments in the laboratory have supported the validity of the Darcy-based flow theory—others have not. Similarly, some observations in the field agree fairly well with the predictions of the flow theory—others do not. When the list of complications that can be encountered in field soils (and also in laboratory flow columns) is considered, it is not surprising that deviations from theory are found, and perhaps it is more surprising that the theory works as well as it does on occasion. Almost any of the complications identified above can, under the right circumstances, cause significant and even drastic deviations from the Darcy-based flow theory.

However, as a first approximation (and sometimes a very good one) the theory embodied in equations [2] and [3] is sufficiently valid for predicting unsaturated flow behavior. Even when it is not a good model, it provides a point of departure for a modified or extended theory to treat more complicated systems.

The growing body of analyses (analytic, quasi-analytic, approximate, and numerical) that is appearing on the literature provides a "catalogue" of solutions or descriptions of behavior of a set of limiting and/or ideal cases. It is not likely that a given field flow situation will be precisely identifiable with any one of these ideal cases. However, there may be an approximate identification of the field flow situation with one or more ideal cases, and with skill and experience in application of the concepts of unsaturated flow, it may be possible to gain a degree of insight into the behavior of the field flow situation.

The advent of high-speed computers has made possible the treatment of complexities in flow system analysis that heretofore could not be attempted. For example, spatial and temporal variability of the hydraulic properties, hysteresis, and gas phase flow can, in principle, be incorporated into an analysis if they can be characterized by appropriate functions or "laws." The importance of the various complexities may be investigated using numerical solution techniques on the flow equation(s) that are believed to model the system. It may be possible, with sufficient experience, to perceive generalities, simplifications, and approximations from such studies that may be applied to field flow situations. However, numerical solutions incorporating complexities, such as variable hydraulic properties and hysteresis, are expensive not only in terms of computational costs, but also in terms of data collection costs required to characterize the complexity. Detailed analyses of this type will probably not be conducted on a routine basis for prediction of soil water flow in the field. Qualitative and approximate methods are probably sufficient for many purposes. In certain specific instances, where the potential economic return or benefit of detailed knowledge of the behavior of a given flow system is sufficiently large, the detailed measurements and solution process may be justified.

The measurement of the hydraulic properties $K(\theta)$ and $\theta(h)$ is still the central problem in applying Darcy-based flow theory. The methods at hand are not particularly rapid. The basic principles for hydraulic conductivity measurement are well enough known—what is needed is improvement of techniques.

LITERATURE CITED

Abd-El-Aziz, M. H., and S. A. Taylor. 1965. Simultaneous flow of water and salt through unsaturated porous media: I. Rate equations. Soil Sci. Soc. Amer. Proc. 29: 141–143.

Anderson, D. M., and P. Hoekstra. 1965. Migration of interlamellar water during freezing and thawing of Wyoming bentonite. Soil Sci. Soc. Amer. Proc. 29:498–503.

Bear, Jacob. 1969. Hydrodynamic dispersion. p. 109–199. *In* Roger J. M. de Wiest (ed.) Flow through porous media. Academic Press, New York.

Bear, Jacob. 1970. Two-liquid flows in porous media. Advan. Hydrosci. 6:142–249.

Biot, M. A. 1956. General solutions of the equations of elasticity and consolidation for a porous material. J. Appl. Mech. 23:91–96.

Black, T. A., W. R. Gardner, and C. B. Tanner. 1970. Water storage and drainage under a row crop on a sandy soil. Agron. J. 62:48–51.

Black, T. A., W. R. Gardner, and G. W. Thurtell. 1969. The prediction of evaporation drainage and soil water storage for a bare soil. Soil Sci. Amer. Proc. 33: 655–660.

Bloomsburg, G. L., and A. T. Corey, 1964. Diffusion of entrapped air from porous media. Colorado State Univ., Fort Collins, Colo., Hydrology Paper 5.

Bruce, R. R., and A. Klute. 1956. The measurement of soil moisture diffusivity. Soil Sci. Soc. Amer. Proc. 20:458–462.

Brust, K. J., C. H. M. van Bavel, and G. B. Stirk. 1968. Hydraulic properties of a clay loam soil and the field measurement of water uptake by roots: III. Comparison of field and laboratory data on retention and of measured and calculated conductivities. Soil Sci. Soc. Amer. Proc. 32:322–326.

Carson, J. E., and H. Moses. 1963. The annual and diurnal heat exchange cycles in the upper layers of the soil. J. Appl. Meteorol. 2:397–406.

Cary, J. W. 1966. Soil moisture transport due to thermal gradients. Soil Sci. Soc. Amer. Proc. 30:428–433.

Cary, J. W., and S. A. Taylor. 1962a. The interaction of the simultaneous diffusions of heat and water vapor. Soil Sci. Soc. Amer. Proc. 26:413–417.

Cary, J. W., and S. A. Taylor. 1962b. Thermally driven liquid and vapor phase transfer of water and energy in soil. Soil Sci. Soc. Amer. Proc. 26:417–420.

Cary, J. W., and S. A. Taylor. 1967. The dynamics of soil water. Part II. Temperature and solute effects. *In* R. M. Hagan et al. (ed.) Irrigation of agricultural lands. Agronomy 11:245–253.

Childs, E. C. 1967. Soil moisture theory. Advan. Hydrosci. 4:73–116.

Childs, E. C. 1969. The physical basis of soil water phenomena. John Wiley & Sons, New York.

Christansen, D. R., and Hayden Ferguson. 1966. The effect of interactions of salts and clays on unsaturated water flow. Soil Sci. Soc. Amer. Proc. 30:549–553.

Corey, A. T., and W. D. Kemper. 1961. Concept of total potential in water and its limitations. Soil Sci. 91:299–302.

Corey, G. L., A. T. Corey, and R. H. Brooks. 1965. Similitude for non-steady drainage of partially saturated soils. Colorado State Univ., Fort Collins, Colo., Hydrology Paper No. 9.

Davidson, J. M., D. R. Nielsen, and J. W. Biggar. 1966. The dependence of soil water uptake and release upon the applied pressure increment. Soil Sci. Soc. Amer. Proc. 30:298–302.

Davidson, J. M., L. R. Stone, D. R. Nielsen, and M. E. LaRue. 1969. Field measurement and use of soil water properties. Water Resour. Res. 6:1312–1321.

Day, P. R., and J. N. Luthin. 1956. A numerical solution of the differential equation of flow for a vertical drainage problem. Soil Sci. Soc. Amer. Proc. 20:443–447.

DeBacker, L. W. 1967. Measurement of entrapped gas in the study of unsaturated flow phenomena. Water Resour. Res. 3:245–249.

Debano, L. F. 1971. Effect of hydrophobic substances on water movement in soils. Soil Sci. Soc. Amer. Proc. 35:340–343.

Debano, L. F., and J. Letey (ed.). 1969. Proceedings symposium on water repellent soils. Univ. Calif., Riverside.

Dirksen, C., and R. D. Miller. 1966. Closed system freezing of unsaturated soil. Soil Sci. Soc. Amer. Proc. 30:168–173.

Doering, E. J. 1965. Soil water diffusivity by the one-step method. Soil Sci. 99:322–326.

Elrick, D. E. 1961. Transient two-phase capillary flow in porous media. Physics of Fluids 4:572–575.

Elrick, D. E., and D. H. Bowman. 1964. Note on an improved apparatus for soil moisture flow measurements. Soil Sci. Soc. Amer. Proc. 28:450–451.

Ferguson, H., P. L. Brown, and D. D. Dickey. 1964. Water movement and loss under frozen soil conditions. Soil Sci. Soc. Amer. Proc. 28:700–703.

Freeze, R. A. 1969. The mechanism of natural ground water recharge and discharge. I. One-dimensional, vertical, unsteady, unsaturated flow above a recharging or discharging ground-water flow system. Water Resour. Res. 5:153–171.

Gardner, W. R. 1956. Calculation of capillary conductivity from pressure plate outflow data. Soil Sci. Soc. Amer. Proc. 20:317–320.

Gardner, W. R. 1959a. Diffusivity of soil water during sorption as affected by temperature. Soil Sci. Soc. Amer. Proc. 23:406–407.

Gardner, W. R. 1959b. Solutions of the flow equation for the drying of soils and other porous media. Soil Sci. Soc. Amer. Proc. 23:183–187.

Gardner, W. R. 1962. Approximate solution of a non-steady state drainage problem. Soil Sci. Soc. Amer. Proc. 26:129–132.

Gardner, W. R., and D. I. Hillel. 1962. The relation of external evaporation conditions to the drying of soils. J. Geophys. Res. 67:4319–4325.

Gardner, W. R., M. S. Mayhugh, J. O. Goertzen, and C. A. Bower. 1959. Effect of electrolyte concentration and exchangeable sodium percentage on diffusivity of water in soils. Soil Sci. 88:270–274.

Garner, Dale E., John K. Donaldson, and George S. Taylor. 1969. Entrapped soil air in a field site. Soil Sci. Soc. Amer. Proc. 33:634–635.

Green, Don W., Hassan Dabiri, Charles F. Weinaug, and Robert Prill. 1970. Numerical modeling of unsaturated ground water flow and comparison of the model to a field experiment. Water Resour. Res. 6:862–874.

Green, R. E., and J. C. Corey. 1971. Calculation of hydraulic conductivity: A further evaluation of some predictive methods. Soil Sci. Soc. Amer. Proc. 35:3–8.

Green, R. E., R. J. Hanks, and W. E. Larson. 1964. Estimates of field infiltration by numerical solution of the moisture flow equation. Soil Sci. Soc. Amer. Proc. 28: 15–19.

Green, W. H., and G. A. Ampt. 1911. Studies on soil physics: I. Flow of air and water through soils. J. Agr. Sci. 4:1–24.

Groenevelt, P. H., and G. H. Bolt. 1969. Non-equilibrium thermodynamics of the soil-water system. J. Hydrol. 7:358–388.

Hanks, R. J., and S. A. Bowers. 1962. Numerical solution of the moisture flow equation for infiltration into layered soils. Soil Sci. Soc. Amer. Proc. 26:530–534.

Hanks, R. J., A. Klute, and E. Bresler. 1969. A numeric method for estimating infiltration redistribution drainage and evaporation of water from soil. Water Resour. Res. 5:1064–1069.

Hillel, D. 1970. Soil and water. Academic Press, New York.

Hoekstra, Pieter. 1966. Moisture movement in soils under temperature gradients with the cold side temperature below freezing. Water Resour. Res. 2:241–250.

Ibrahim, H. A., and W. Brutsaert. 1968. Intermittent infiltration into soils with hysteresis. Amer. Soc. Civil Eng. J. Hydraul. Div. HY1:113–137.

Jackson, R. D. 1963. Temperature and soil water diffusivity relations. Soil Sci. Soc. Amer. Proc. 27:363–366.

Jackson, R. D. 1964a. Water vapor diffusion in relatively dry soil. I. Theoretical considerations and sorption experiments. Soil Sci. Soc. Amer. Proc. 28:172–176.

Jackson, R. D. 1964b. Water vapor diffusion in relatively dry soil. III. Steady-state experiments. Soil Sci. Soc. Amer. Proc. 28:467–470.

Jackson, R. D., R. J. Reginato, and C. H. M. van Bavel. 1965. Comparison of measured and calculated hydraulic conductivities of unsaturated soil. Water Resour. Res. 1: 375–380.

Jackson, R. D., and F. D. Whisler. 1970. Equations for approximating vertical non-steady state drainage of soil columns. Soil Sci. Soc. Amer. Proc. 34:715–718.

Jensen, R. D., M. Haridasan, and G. S. Rahi. 1970. The effect of temperature on water flow in soils. Water Resour. Res. Institute, Mississippi State Univ., State College, Miss.

Kemper, W. D. 1961. Movement of water as affected by free energy and pressure gradients: II. Experimental analysis of porous systems in which free energy and pressure gradients act in opposite directions. Soil Sci. Soc. Amer. Proc. 25:260–265.

Kemper, W. D., and J. B. Rollins. 1966. Osmotic efficiency coefficients across compacted clays. Soil Sci. Soc. Amer. Proc. 30:528–534.

Klute, A. 1952. A numerical method for solving the flow equation for water in unsaturated materials. Soil Sci. 73:105–116.

Klute, A. 1965. Water capacity. *In* C. A. Black et al. (ed.) Methods of soil analysis, Part I. Agronomy 9:273–278.

Klute, A. 1969. The flow of water in unsaturated soil. *In* The progress of hydrology. Int. Seminar for Hydrology Professors, Proc. 1st. Urbana, Ill. II:821–886.

LaRue, M. E., D. R. Nielsen, and R. M. Hagan. 1968. Soil water flux below a ryegrass root zone. Agron. J. 60:625–629.

Letey, J. 1968. Movement of water through soil as influenced by osmotic pressure and temperature gradients. Hilgardia 39:405–418.

Letey, J., W. D. Kemper, and L. Noonan. 1969. The effect of osmotic pressure gradients on water movement in unsaturated soil, Soil Sci. Soc. Amer. Proc. 33:15–18.

Marshall, T. J. 1958. A relation between permeability and size distribution of pores. J. Soil Sci. 9:1–8.

McNeal, B. L., and N. T. Coleman. 1966. Effect of solution composition on soil hydraulic conductivity. Soil Sci. Soc. Amer. Proc. 30:308–312.

Miller, D. E., and W. H. Gardner. 1962. Water infiltration into stratified soil. Soil Sci. Soc. Amer. Proc. 26:115–118.

Miller, E. E., and D. E. Elrick. 1958. Dynamic determination of capillary conductivity extended for non-negligible membrane impedance. Soil Sci. Soc. Amer. Proc. 22: 483–486.

Miller, E. E., and A. Klute. 1967. The dynamics of soil water. Part I. Mechanical forces. *In* R. M. Hagen et al. (ed.) Irrigation of agricultural lands. Agronomy 11:209–240.

Miller, E. E., and R. D. Miller. 1956. Physical theory for capillary flow phenomena. J. Appl. Phys. 27:324–332.

Millington, R. J., and J. P. Quirk. 1959. Permeability of porous media. Nature 183: 387–388.

Millington, R. J., and J. P. Quirk. 1961. Permeability of porous solids. Trans. Faraday Soc. 57:1200–1207.

Nielsen, D. R., J. M. Davidson, J. W. Biggar, and R. J. Miller. 1964. Water movement through Panoche clay loam soil. Hilgardia 35:491–506.

Nielsen, D. R., R. D. Jackson, J. W. Cary, D. D. Evans (ed.) 1972. Soil water. American Society of Agronomy and Soil Science Society of America, Madison, Wis. 176 p.

Nielsen, D. R., Don Kirkham, and E. R. Perrier. 1960. Soil capillary conductivity: Comparison of measured and calculated values. Soil Sci. Soc. Amer. Proc. 24:157–160.

Nielsen, D. R., Don Kirkham, and W. R. Van Wijk. 1961. Diffusion equation calculations of field soil water infiltration profiles. Soil Sci. Soc. Amer. Proc. 25:165–168.

Ogata, G., and L. A. Richards. 1957. Water content changes following irrigation of bare-field soil that is protected from evaporation. Soil Sci. Soc. Amer. Proc. 21:355–356.

Peck, A. J. 1960. Change of moisture tension with temperature and air pressure: theoretical. Soil Sci. 89:303–310.

Peck, A. J. 1965. Moisture profile development and air compression during water by bounded porous bodies. 3. Vertical columns. Soil Sci. 100:44–52.

Philip, J. R. 1955. Numerical solution of equations of the diffusion type with diffusivity concentration dependent. Trans. Faraday Soc. 51:885–892.

Philip, J. R. 1957a. Numerical solution of equations of the diffusion type with diffusivity concentration dependent. II. Aust. J. Physics 10:29–42.

Philip, J. R. 1957b. Evaporation and moisture and heat fields in the soil. J. Meteorol. 14:354–366.

Philip, J. R. 1964. Similarity hypothesis for capillary hysteresis in porous materials. J. Geophys. Res. 69:1553–1562.

Philip, J. R. 1967. Sorption and infiltration in heterogeneous media. Aust. J. Soil Res. 5:1–10.

Philip, J. R. 1968a. The theory of absorption into aggregated media. Aust. J. Soil Res. 6:1–19.

Philip, J. R. 1968b. Diffusion, dead-end pores and linearized absorption in aggregated media. Aust. J. Soil Res. 6:21–30.

Philip, J. R. 1969a. Hydrostatics and hydrodynamics in swelling soils. Water Resour. Res. 5:1070–1077.

Philip, J. R. 1969b. Theory of infiltration. Advan. Hydrosci. 5:216–219.

Philip, J. R. 1970. Flow in porous media. Annu. Rev. Fluid Mech. 2:177–204.

Philip, J. R., and D. A. deVries. 1957. Moisture movement in porous materials under temperature gradients. Trans. Amer. Geophys. Union 38:222–232.

Philip, J. R., and D. E. Smiles. 1969. Kinetics of sorption and volume change in three-component systems. Aust. J. Soil Res. 7:1–19.

Poulovassilis, A. 1962. Hysteresis of pore water: an application of the concept of independent domains. Soil Sci. 93:405–412.

Poulovassilis, A. 1969. The effect of hysteresis of pore-water on the hydraulic conductivity. J. Soil Sci. 20:52–56.

Quirk, J. P., and R. K. Schofield. 1955. The effect of electrolyte concentration on soil permeability. J. Soil Sci. 6:163–178.

Reeve, R. C. 1960. The transmission of water by soils as influenced by chemical and physical properties. Int. Congr. Agr. Eng., Trans. 5th (Brussels, Belgium) I:21–32.

Rhoades, J. D., and R. D. Ingvalson. 1969. Macroscopic swelling and hydraulic conductivity properties of four vermiculitic soils. Soil Sci. Soc. Amer. Proc. 33:364–369.

Richards, L. A., and D. C. Moore. 1952. Influence of capillary conductivity and depth of wetting on moisture retention in soil. Trans. Amer. Geophys. Union 33:531–540.

Richards, L. A., W. R. Gardner, and G. Ogata. 1956. Physical processes determining water loss from soil. Soil Sci. Soc. Amer. Proc. 20:310–314.

Rose, C. W. 1968a. Water transport in soil with a daily temperature wave. I. Theory and experiment. Aust. J. Soil Res. 6:31–44.

Rose, C. W. 1968b. Water transport in soil with a daily temperature wave. II. Analysis. Aust. J. Soil Res. 6:45–57.

Rose, C. W., W. R. Stern, and J. E. Drummond. 1965. Determination of hydraulic conductivity as a function of depth and water content for soil *in situ*. Aust. J. Soil Res. 3:1–9.

Rose, D. A. 1963a. Water movement in porous materials. Part I. Isothermal vapour transfer. Brit. J. Appl. Phys. 14:256-262.

Rose, D. A. 1963b. Water movement in porous materials. Part II. The separation of the components of water movement. Brit. J. Appl. Phys. 14:491-496.

Rubin, J. 1967. Numerical method for analyzing hysteresis-affected post infiltration redistribution of soil moisture. Soil Sci. Soc. Amer. Proc. 31:13-20.

Rubin, J., and R. Steinhardt. 1963. Soil water relations during rain infiltration: I. Theory. Soil Sci. Soc. Amer. Proc. 27:246-250.

Rubin, J., and R. Steinhardt. 1964. Soil water relations during rain infiltration: III. Water uptake at incipient ponding. Soil Sci. Soc. Amer. Proc. 28:614-620.

Stallman, R. W. 1967. Flow in the zone of aeration. Advan. Hydrosci. 4:151-195.

Staple, W. J. 1969. Comparison of computed and measured moisture redistribution following infiltration. Soil Sci. Soc. Amer. Proc. 33:840-847.

Swartzendruber, Dale. 1966. Soil-water behavior as described by transport coefficients and functions. Advan. Agron. 18:327-362.

Tanner, C. B., and D. E. Elrick. 1958. Volumetric porous pressure plate apparatus for moisture hysteresis measurements. Soil Sci. Soc. Amer. Proc. 22:575-576.

Taylor, S. A., and J. W. Cary. 1960. Analysis of the simultaneous flow of water and heat or electricity with the thermodynamics of irreversible processes. Int. Congr. Soil Sci., Trans. 7th (Madison, Wis.) 1:80-90.

Taylor, S. A., and J. W. Cary. 1964. Linear equations for the simultaneous flow of matter and energy in a continuous soil system. Soil Sci. Soc. Amer. Proc. 28: 167-172.

Terzaghi, K. 1943. Theoretical soil mechanics. John Wiley & Sons, New York.

Topp, G. C. 1969. Soil-water hysteresis measured in a sandy loam and compared with the hysteretic domain model. Soil Sci. Soc. Amer. Proc. 33:645-651.

Topp, G. C. 1969. Soil water hysteresis: the domain theory extended to pore interaction conditions. Soil Sci. Soc. Amer. Proc. 35:219-225.

Topp, G. C., and E. E. Miller. 1966. Hysteretic moisture characteristics and hydraulic conductivities for glass bead media. Soil Sci. Soc. Amer. Proc. 30:156-162.

Topp, G. C., A. Klute, and D. B. Peters. 1967. Comparison of water content-pressure head data obtained by equilibrium, steady-state and unsteady-state methods. Soil Sci. Soc. Amer. Proc. 31:312-314.

van Bavel, C. H. M., G. B. Stirk, and K. J. Brust. 1968. Hydraulic properties of a clay loam soil and the field measurement of water uptake by roots: I. Interpretation of water content and pressure profiles. Soil Sci. Soc. Amer. Proc. 32:310-317.

Wang, F. C., and V. Lakshiminarayana. 1968. Mathematical simulation of water movement through unsaturated non-homogeneous soils. Soil Sci. Soc. Amer. Proc. 32: 329-334.

Watson, C. L., and J. Letey. 1970. Indices for characterizing soil water repellancy based upon contact angle surface tension relationships. Soil Sci. Soc. Amer. Proc. 34: 841-844.

Whisler, F. D., and A. Klute. 1965. The numerical analysis of infiltration considering hysteresis, into a vertical soil column at equilibrium under gravity. Soil Sci. Soc. Amer. Proc. 29:489-494.

Whisler, F. D., and A. Klute. 1966. Analysis of infiltration into stratified soil columns. p. 451-470. *In* P. E. Ribtema (ed.) Proc. Symposium on Water in the Unsaturated Zone. UNESCO and Govt. of Netherlands, Wageningen.

Wilkinson, G. E., and A. Klute. 1962. The temperature effect on equilibrium energy status of water held by porous media. Soil Sci. Soc. Amer. Proc. 26:326-329.

Williams, P. J. 1968. Properties and behavior of freezing soils. Nat. Res. Counc. Can., Div. Bldg. Res., Ottawa, Ont. Research Paper no. 359.

Willis, W. O., H. L. Parkinson, C. W. Carlson, and H. J. Haas. 1964. Water table changes and soil moisture loss under frozen conditions. Soil Sci. 98:244-248.

Wilson, L. G., and J. N. Luthin. 1963. Effect of airflow ahead of the wetting front on infiltration. Soil Sci. 96:136-143.

Yaron, Bruno, and G. W. Thomas. 1968. Soil hydraulic conductivity as affected by sodic water. Water Resour. Res. 4:545-552.

Youngs, E. G. 1960. The drainage of liquids from porous materials. J. Geophys. Res. 65:4025-4030.

Diurnal Changes in Soil Water Content During Drying[1]

3

RAY D. JACKSON[2]

ABSTRACT

In the surface zone of a field soil, the soil water content exhibits a marked diurnal variation. The surface dries during the day and partially rewets at night. This variation plays a significant, but little understood role in the evaporation of water from soil. In two experiments, soil-water contents to 9 cm were measured in 1-cm increments at 0.5-hour intervals for 16 days during March 1971, and at 1-hour intervals for 7 days during July 1970. Diurnal soil water content changes, and soil water content and soil water flux profiles at several times are compared for the two seasons. The effect of the rate of drying on cumulative evaporation is discussed. The soil water regime under natural environmental conditions was so dynamic that the three stages of drying could not be delineated.

INTRODUCTION

The surface layer of soil is of utmost importance to man. He cultivates this layer and plants the seeds from which crops grow. The maintenance of an optimum environment for seeds and plants has been the goal of farmers for centuries. A key ingredient in this environment is water. Various management practices have been devised to control the water content of the seedbed. In areas of the world where water is not plentiful, evaporation from the soil surface is a major concern. A better understanding of the evaporation process requires detailed knowledge of the water regime of the surface zone and the factors that influence it.

In a natural environment, the soil surface undergoes diurnal changes in water content and temperature. Atmospheric variables such as radiation, wind, air temperature, and humidity influence the physical condition of the soil surface and determine the course of evaporation. The rate of water movement to the surface at a particular time is determined by the ability of the soil to conduct water, the evaporative demand at the surface, the temperature of the soil-water system, and the soil water pressure-, temperature-, and salt-gradients. These various factors change with time. Some, notably the temperature and temperature gradient, change markedly during the course of a day.

---

[1] Contribution from the Western Region, Agricultural Research Service, U. S. Department of Agriculture, Phoenix, Arizona.
[2] Research Physicist, U. S. Water Conservation Laboratory, Phoenix, Arizona.

Classical theories of soil water movement have, for the most part, ignored the effects of variable temperature and temperature gradient. During the past two decades, the theories of simultaneous movement of heat and water and the thermodynamics of the irreversible processes have been applied with some success to the movement of soil water in nonisothermal systems. Except for the experiments of Rose (1968a, 1968b, 1968c) and the calculations of Cary (1966), tests of these theories have been largely restricted to the laboratory. Rose measured water content and temperature profiles in the 0- to 15-cm layer 6 to 10 times a day for 6 days. His results clearly demonstrate the dynamic nature of the soil water system. Lacking a direct measurement, he calculated evaporation using the theory of simultaneous movement of heat and water.

This paper presents an overview of two field experiments, carried out during two different seasons, in which soil water content profiles were measured at 0.5-hour and 1-hour intervals for periods of 1 to 2 weeks after irrigation. Concurrently, evaporation rates, soil temperatures, and pertinent meteorological parameters were measured. The main emphasis here concerns the description of diurnal soil water content changes, soil water content profiles, and soil water flux at several depths and times. It is anticipated that subsequent publications will apply theories of simultaneous flow of heat and water to these data, and further elucidate the role of meteorological parameters on the evaporation of water from bare soil.

EXPERIMENTAL PROCEDURES

The experimental site was a 72 by 90-m field at the U. S. Water Conservation Laboratory, Phoenix, Arizona. Three weighing lysimeters located within the field have been described by Van Bavel, Fritschen, and Reginato (1963). The soil, Adelanto loam, is reasonably uniform to about 100 cm and has been cultivated numerous times during past years. The field was divided into three plots, each plot surrounding a lysimeter. At the start of each experiment two of the lysimeters and the surrounding plots were irrigated with about 10 cm of water. After irrigation the lysimeter weight loss and hence evaporation was monitored at 0.5-hour intervals throughout the experiment.

The first experiment was conducted from 10–17 July 1970. The field and lysimeters were irrigated on the afternoon of 10 July. These 7 days were characterized by clear skies and light winds, except from 0330 to 0600 on 16 July when winds were gusty. Maximum and minimum air temperatures ranged from 40 to 44C and 17 to 30C, respectively.

The second, and more extensive experiment, was conducted during March and April 1971. On the afternoon of 2 March the field and lysimeters were irrigated. Soil water content measurements were made from 5–18 March, 25 March, and 8 April. Generally, skies were clear and winds were light. Maximum and minimum air temperatures ranged from 17 to 24C and −2 to 5C, respectively. High winds and cloudy skies prevailed on 5 and 13 March.

Soil water contents were measured gravimetrically. In July, soil cores were taken to 10 cm and sectioned into 1-cm increments. Five sites were sampled and the depth increments composited. A separate sample was taken for the 0- to 0.5-cm increment. Samples were taken at 1-hour intervals from 0500 on 11 July to 2200 on 17 July.

During March 1971, samples were taken of the 0- to 0.5-cm increment, in 1-cm increments to 5 cm, and in 2-cm increments from 5 to 9 cm. Samples were taken at 0.5-hour intervals from 2300 on 4 March 1971 to 0130 on 19 March 1971, 2300 on 24 March to 0130 on 26 March, and 2300 on 7 April to 0130 on 9 April. Six sites were sampled each time and composited for each depth increment. In both experiments, the time between the beginning of sampling until the samples were weighed and placed in the oven (at 110C) was less than 15 min.

Errors inherent in the gravimetric measurement of soil water content

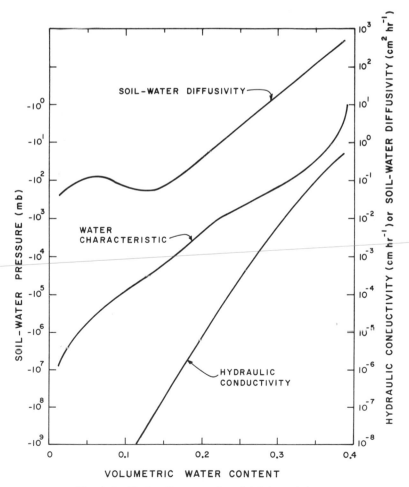

Figure 1. Hydraulic properties of Adelanto loam.

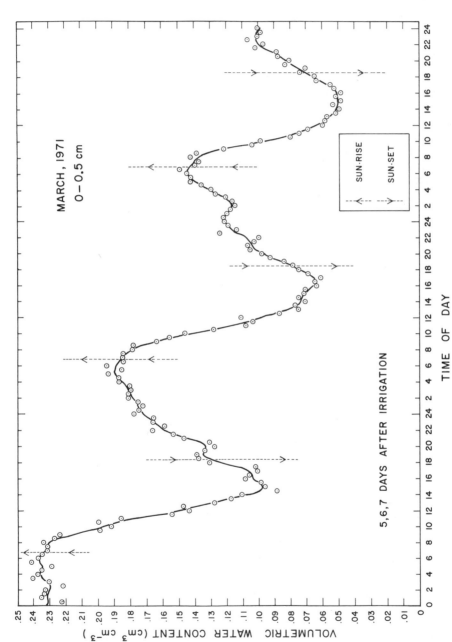

Figure 2. Volumetric water content in the 0- to 0.5-cm increment versus time for 3 days during March 1971. The solid line represents smoothed data and the symbols represent the measured values.

arise from site differences, sectioning of cores, weighing, and recording data. Site differences and sectioning errors were minimized by compositing samples from six sites. All weighings were made to 0.01 g on a direct-reading balance. To reduce scatter, the water content data were smoothed by a 1-2-3-2-1 weighted running average procedure. Water contents were converted to the volumetric basis by multiplying by the bulk density. The bulk densities used were the average of five core samples for each depth taken on 5 different days.

During the July experiment, thermocouples were placed at the soil surface and at 1, 2, 4, 8, 16, 32, 64, and 128 cm. In March temperatures at depths 0.5, 3, and 5 cm were also measured. Other measurements were solar and net radiation, reflected solar radiation, windspeed at several heights above the surface, and vapor pressure and air temperature at two heights above the soil. All meteorological data and soil temperatures were recorded on punched paper tape by a data acquisition system.

Several physical parameters of Adelanto loam (the water characteristic, the hydraulic conductivity, and the soil water diffusivity) are presented in Fig. 1. The water characteristic for Adelanto loam is comprised of field data taken from Brust, Van Bavel, and Stirk (1968) and, for the lower water contents, laboratory data of Jackson (1964b).

The hydraulic conductivities were calculated from the water characteristic using the procedure of Millington and Quirk (1959) as used by Jackson, Reginato, and Van Bavel (1965). The matching factor was obtained from the hydraulic conductivity data of Brust et al. (1968). Soil water diffusivities at the higher water contents were calculated from the hydraulic conductivity and water characteristic data. Below 0.1 water content, the curve represents the measured vapor diffusion coefficients of Jackson (1964a).

RESULTS AND DISCUSSION

Diurnal Soil Water Content Changes

Figure 2 depicts the soil water content in the 0- to 0.5-cm layer at 0.5-hour intervals during March 1971. These data show the wide range of water contents encountered in this layer during the course of a day. Water content decreased 50% between 0600 and 1400 on 7 March. Following recovery of volumetric water content from 0.097 to 0.19 during the night, this layer again dried rapidly the next day. This pattern was exhibited by the 0- to 0.5-cm layer for all days of measurement. Water loss began about 1 to 2 hours before sunrise and rewetting began about 2 to 4 hours before sunset.

Figure 2 also shows the results of the smoothing procedure. The circles represent the measured volumetric water content (obtained from multiplying the measured gravimetric data by the bulk density), whereas the solid line connects the points calculated by the smoothing procedure. Only smoothed data will be presented in subsequent figures.

The amplitude of the diurnal water content change decreases with depth as shown in Fig. 3. Five days' data are presented to show the differences in diurnal patterns as the soil dried. For day 3 after irrigation, water contents are shown for three depths. Data for the other five increments lay between the extremes shown. In the 0- to 0.5-cm increment, the water content was 0.35 before sunrise and decreased to 0.24 at sunset. Recovery began at sunset and increased to 0.295 the following morning (data for day 4 are not shown). At the 7- to 9-cm depth the water content decreased from 0.305 to 0.275 during the day, but showed only a small recovery at night.

By day 9 (11 March 1971), the water content in the 0- to 0.5-cm layer has decreased considerably, exhibiting a range of 0.093 to 0.038 with the maximum occurring near sunrise and the minimum about 2 hours before sunset. Although the amplitude decreases with time, the diurnal change is evident for every measurement day including the 37th and final day of the experiment.

The lines for 0.5- to 1-cm were calculated from the data for 0- to 0.5-cm and 0- to 1-cm increments. Errors in measurement at the two depths, not removed in smoothing, can be magnified in the calculation. Nevertheless, the calculated lines correspond very well with the other data. The minimum water content for the 0.5- to 1-cm layer lagged behind that of the 0- to 0.5-cm layer for four of the days shown in Fig. 3. For day 9 the minimum was near sunset, about 2 hours later than the minimum for 0- to 0.5-cm. On subsequent days the minimum occurred earlier, and at day 37 occurred at nearly the same time as for the 0- to 0.5-cm layer. No time lags were evident on day 3.

The amplitude of the diurnal fluctuation decreased with depth and time. On day 9 the 1- to 2-cm layer dried during the afternoon and rewet at night, whereas at deeper depths fluctuations occurred at various times during the day. The water content increase at these depths near midday may be due to downward flow from the shallower layers caused by temperature gradients. However, the deeper layers exhibited a net decrease in water content from midnight to midnight. The range in water content changes at the 7- to 9-cm layer was 0.008 on day 9, compared to 0.03 on day 3.

Soil Water Flux

Evaporation rates vary with the season of the year; hence, water content and flux profiles within the surface zone should be different for the different seasons. In this and the following section, evaporation rates, cumulative evaporation, water content, and flux profiles are compared for two seasons.

The evaporation rate (cm day^{-1}) for the July 1970 and March 1971 experiments are presented in Fig. 4. Also shown are the soil water flux data at the 9-cm depth. The flux at 9 cm was calculated from the flux at the surface (obtained from the lysimeters) and the water content changes in the 0- to 9-cm layer. For both experiments the flux per 24 hours at 9 cm was upward

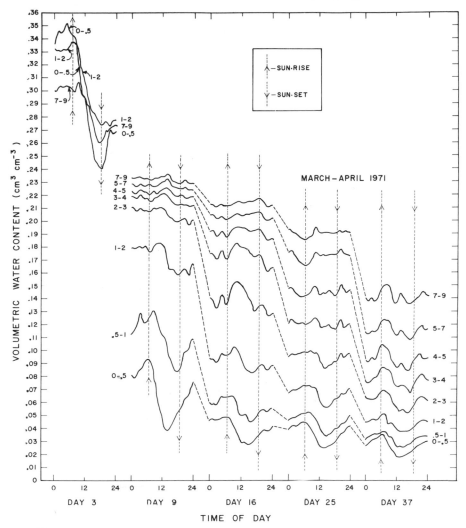

Figure 3. Volumetric water content versus time for several depths at 3, 9, 16, 23, and 37 days after irrigation.

towards the evaporating surface. For day 11 (13 March), the flux at 9 cm was almost equal to the evaporation rate at the surface. This indicates essentially no net loss of water from the 0- to 9-cm layer for this day. Day 11 was characterized by high gusty winds during the afternoon and evening. The average windspeed for the 24-hour period was more than twice that for the surrounding days. For the July experiment, the flux at 9 cm was almost equal to the evaporation rate on day 6. Between 0300 and 0600 hours high gusty winds prevailed. In both cases the water contents of the surface few centimeters were relatively low, thus much of the pore space was available for air and water vapor movement. It is possible that mass flow of

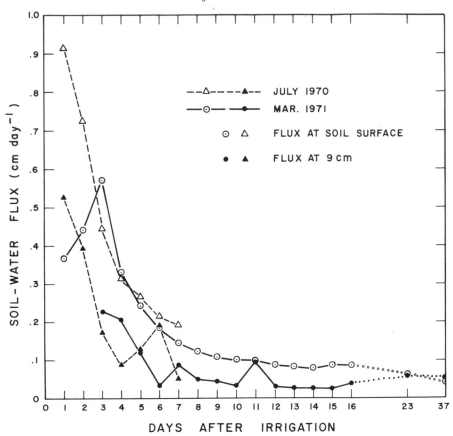

Figure 4. Daily evaporation at the soil surface and soil water flux at 9 cm.

water vapor due to wind-induced air pressure fluctuations contributed significantly to the net upward flux during this period (Kimball & Lemon, 1971).

During the last days of the experiment, the flux at 9 cm approached the evaporation rate. On day 23, they were nearly equal, while on day 37, the flux at 9 cm was slightly higher than at the surface. This latter condition would usually not hold for additional days, since the soil continues to dry. At the end of 7 days in July, 69% of the total water evaporated originated in the 0- to 9-cm layer. For 16 days in March, 66% came from the 0- to 9-cm layer.

Figure 5 shows the cumulative water loss at the surface and the cumulative flux at 9 cm. More water was lost by evaporation from the entire soil profile during 7 days in July than during 16 days in March. Also, more water was lost from the 0- to 9-cm layer during the 7 July days than the 16 March days. On day 1 of the July experiment, more water was lost in the 0- to 9-cm layer than was evaporated, indicating drainage occurred below 9 cm for the time immediately following irrigation.

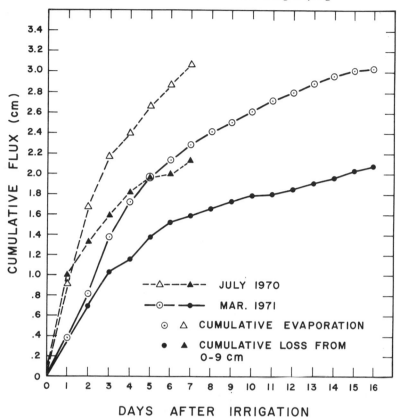

Figure 5. Cumulative evaporation and cumulative loss from the 0-9 cm profile.

Soil Water Content and Flux Profiles

Figure 4 shows that the evaporation rates on day 4 for the July and March experiments were nearly the same. Soil water flux profiles were calculated for these 2 days and are presented in Fig. 6. Profiles for four 6-hour periods were calculated for both days. For these calculations, flow downward was taken as positive. The July data, represented by dashed lines, indicate flow was upward except for the 9-cm depth during the early morning hours (open circles). The flux upward increased in magnitude with time until 1800. For the period 1800-2400, the flux above 3 cm was less in magnitude than for the daylight period, but was greater below 3 cm.

For the March data, a wide range of flux values occurred. The values for the periods 0000-0600 and 1800-2400 indicate water was moving up at all depths between 0 and 9 cm and the water content was increasing at all depths. For the 0600-1200 period, the flux was upward above 3 cm, and downward below 3.5 cm, indicating flow below 9 cm. By the 1200-1800

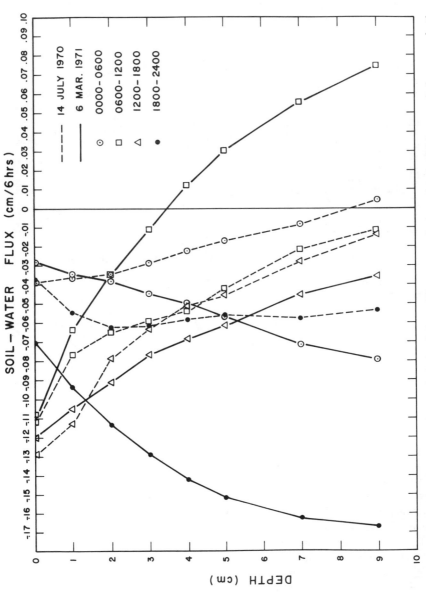

Figure 6. Soil water flux profiles for 4 time periods for 2 days. The evaporation rate for the 24-hour period was nearly the same for the 2 days.

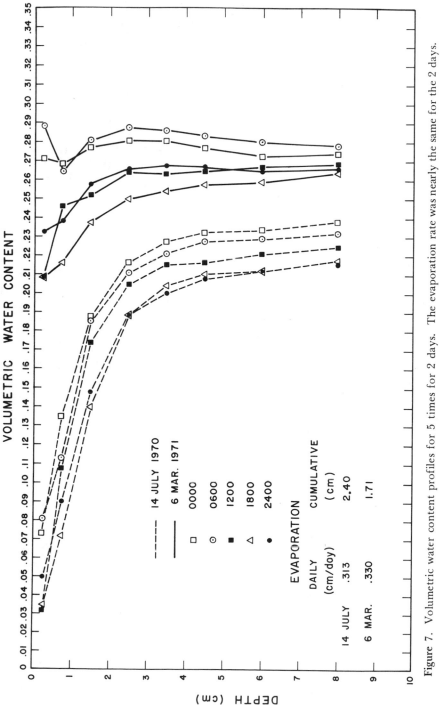

Figure 7. Volumetric water content profiles for 5 times for 2 days. The evaporation rate was nearly the same for the 2 days.

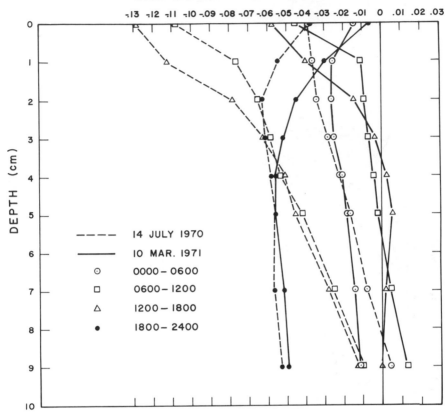

Figure 8. Soil water flux profiles for 4 time periods for 2 days. The cumulative evaporation was nearly the same for the 2 days.

period, all flow was upward with more being lost in the upper layers than
was being replenished at night. The downward flux during 0600–1200 probably was caused by the relatively large temperature gradients that occurred
during this time.

The water content profiles for these 2 days are shown in Fig. 7. It is
obvious that, even though the daily evaporation rates for the 2 days were
nearly equal, the water contents for 6 March were considerably greater than
for 14 July. Five time periods are shown for each day. For July the water
content at all depths steadily decreased with time until 1800. A small increase can be observed for the period 1800–2400. In March there was much
more water content fluctuation with time. The water content decreased
until 1800, but by 2400 had increased to about the 1200 value.

Figure 5 shows that day 4 for July (14th) and day 8 for March (10th)
had nearly the same cumulative evaporation. The flux profiles for four time
periods for these 2 days are compared in Fig. 8. The profiles for 10 March
show less spread than the 6 March data shown in Fig. 6, and they were more

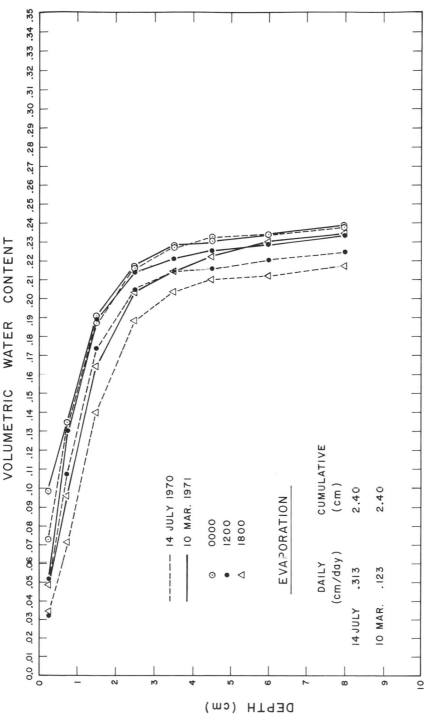

Figure 9. Volumetric water content profiles for 3 times for 2 days during which cumulative evaporation was nearly the same.

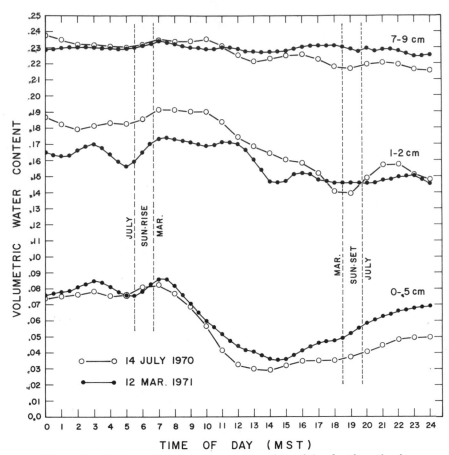

Figure 10. Volumetric water content versus time of day for three depths. Data are for 14 July 1970 and 12 March 1971, 4 and 10 days after irrigation, respectively. The cumulative loss from the 0–9 cm profile was the same for the 2 days.

similar to the 14 July data. The profiles for the time periods 0000–0600 and 1800–2400 were quite similar for the 2 days. The profiles for the daylight periods for March showed less flux upward than those for July.

The similarity of the water content profiles for 14 July and 10 March can be seen in Fig. 9. To preserve clarity only three time periods are shown for each day. The profiles for 0000 hours for the 2 days are nearly identical. For 14 July the profile becomes drier with time of day as reflected in the higher evaporation rate.

A comparison of diurnal water content changes for three depth increments for 2 days is shown in Fig. 10. Data in Fig. 5 indicate that the cumulative loss from the 0- to 9-cm layer was nearly the same on 14 July and 12 March, the 4th and 10th days after irrigation, respectively. The values of water content and the diurnal changes were remarkably similar for the 2 days, especially for the 0- to 0.5-cm and 7- to 9-cm increments. For the 1- to 2-cm increment, the July data were about 0.02 higher in water content for the first 12 hours. However, the evaporation rate for 14 July was about 4 times that of 12 March.

Effect of a Dry Surface Layer

Buckingham (1907) suggested that, if a moist soil is exposed to initially high evaporative conditions, the dry layer that rapidly forms will reduce subsequent evaporation and the cumulative evaporation may in the long run be less than that for a soil initially exposed to low evaporative conditions. This hypothesis has been confirmed by laboratory data (Hillel, 1968). The data presented here allows it to be tested under field conditions.

Water contents in the 0- to 0.5-cm layer for 14 July, 4 days after irrigation (Fig. 7), are in the range 0.03 to 0.08. These water contents correspond to a soil-water pressure range of about -2×10^3 to -2×10^2 bars (Fig. 1), which is indicative of a relatively dry surface layer. This water content range was reached in the March experiment on 10 March, 8 days after irrigation (Fig. 9). Thus a "dry" layer was formed much sooner after irrigation in July than in March, yet subsequently both the cumulative and daily evaporation were higher for July than March. Evaporation data (not shown) for days 8 through 14 in July were also higher than for the March experiment. These data indicate that the early formation of a dry surface layer does not reduce the cumulative evaporation below that which occurs when the surface is moist over a longer period of time.

It can be argued that since the potential evaporation for March was less than for July these results would be expected, and no conclusion can be drawn concerning comparative evaporative losses for the two time periods. However, in the natural environment any condition that would cause the surface soil to remain at a relatively high water content for a time period longer than normal must also be a low potential evaporation condition. Thus, during the same time of year, the hypothesis of a rapidly formed dry layer reducing evaporation losses would not necessarily hold under natural conditions. However, it may hold in cases where the soil has been artificially mulched or cultivated to create a loose dry surface layer.

The Three Stages of Drying

Classically, soil drying has been divided into three stages (Lemon, 1956; Philip, 1957). These are: (i) when the soil is sufficiently moist, evaporation is determined solely by atmospheric conditions; (ii) at intermediate water contents, the ability of the soil to conduct water is the limiting factor with atmospheric conditions no longer as important; and (iii) the third stage is characterized by extremely slow water movement and is dominated by adsorptive

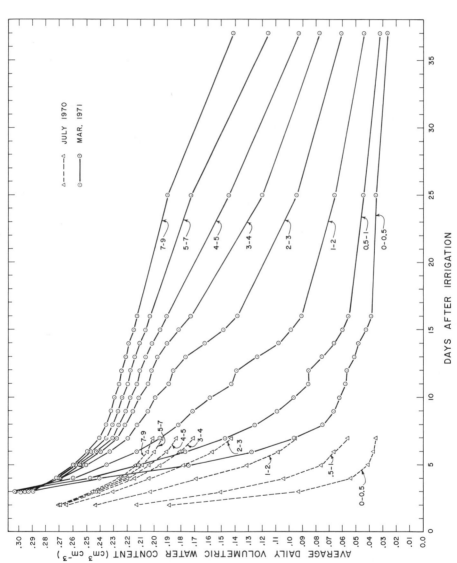

Figure 11. Average daily water content versus days after irrigation for the July 1970 and March 1971 experiments.

forces acting over molecular distances at the solid-liquid interfaces in the soil. As with Buckingham's hypothesis (1907), discussed in the previous section, the concept of the three stages of drying has been demonstrated by laboratory experiments. Their applicability to evaporation under natural conditions is of interest here.

Figure 4 shows that the evaporation rate for July decreased continuously from the first day after irrigation. During March the rate increased for 3 days after irrigation and then decreased continuously, having nearly the same values as the July data. The high winds prevalent on day 3 in March were probably responsible for the high evaporation rate on that day. Assuming that the rate would have been somewhat lower if the wind had been moderate, it is tempting to infer that on day 3 for March, the second stage of drying began. However, the water content at which the change from the first to the second stage of drying takes place cannot be determined from these data.

Wendt et al. (1970) pointed out that the third stage of drying begins when the water content of the soil surface becomes "air dry." Water movement at these low water contents is influenced by adsorptive forces. Jackson (1965) determined, for Adelanto loam, that the water content corresponding to a monomolecular layer was about 0.03. Assuming that physical adsorption takes place in the first two molecular layers, adsorptive forces would influence water movement for water contents below 0.06. For the lack of better criteria, it will be assumed that the third stage of drying begins when the surface dries to 0.06 volumetric water content. Figure 11 shows that the average water content of the 0- to 0.5-cm layer was below 0.06 from the fourth day after irrigation in July. If the criterion is correct, then the July data reflect the third stage of drying from day 4 on. Figure 4 shows that the evaporation rates for days 4–7 in July were very close to or slightly higher than for these days in March. However, the data in Fig. 2 indicate that water contents in the 0- to 0.5-cm layer for these March days were always greater than 0.06, except for 3 hours at midday on day 7. The average water content for the 0- to 0.5-cm layer during March was not less than 0.06 until day 11 (Fig. 11). By this time the evaporation rate (Fig. 4) is less than 0.1 cm/day. Thus, the water content at which the transition from the second to the third stage of drying takes place cannot be determined from these data. The concept of the three stages of drying appears to have little meaning under natural field conditions. This conclusion is supported by the data of Kimball and Jackson (1971).

Application of Soil Water Flow Theory

A pertinent question is how well soil water movement under natural environmental conditions can be predicted with our present knowledge of soil-water flow theory. Klute (1973, Chapter 2, this book) has reviewed soil water flow theory as to its applicability to field situations. Assuming that the soil is "salt free" and ignoring hysteresis, Klute's equation [6] can be applied. As he states, the coefficient of the temperature gradient is the sum of the thermal liquid diffusivity $D_{T\ell}$ and a thermal vapor diffusivity D_{Tv}. The coefficient of the water content gradient is the sum of the isothermal liquid diffusivity $D_{\theta\ell}$ and the isothermal vapor diffusivity $D_{\theta v}$. If these four coefficients are accurately known then soil water flux values corresponding to those given in Fig. 6 and 8 can be calculated. Rose (1968a) and Philip and De Vries (1957) describe how $D_{T\ell}$, $D_{\theta\ell}$, D_{Tv}, and $D_{\theta v}$ can be obtained. None of the various measurement or calculation techniques are free from uncertainty. A method is usually considered reliable if $D_{\theta\ell}$ can be estimated to within a factor of 2. However, an error of this magnitude is intolerable if one is to calculate the values of soil water flux shown in Fig. 6 and 8. For example, if the estimate of $D_{\theta\ell}$ is a factor of 2 greater than the true value (all other terms being accurately known), a calculated flux near the surface would indicate an evaporation rate nearly twice as large as the measured rate.

Uncertainties also exist in the estimation of $D_{\theta v}$, D_{Tv}, and $D_{T\ell}$. All of the coefficients are dependent upon water content and temperature. The temperature dependence of the vapor components are different than the liquid components. These temperature dependencies can be calculated from theory but little experimental verification is available. Perhaps even greater uncertainties exist in the "enhancement factors" discussed by Philip and De Vries (1957) and Rose (1968a, 1968b). These factors were proposed to account for the fact that water vapor movement through soil has been observed (in laboratory studies) to be considerably greater than that predicted by simple diffusion theory. The factors are also temperature and water content dependent. With constantly changing temperature conditions in the natural environment, information concerning the temperature dependence of these factors is essential if accurate predictions are to be made using soil water flow theory.

Obviously, direct application of soil water flow theory to describe diurnal soil water movement near the soil surface is complicated by uncertainties. Data from the experiments reported here allow soil water flux to be calculated at several depths and times. These data may lead to more refined estimates of the diffusion coefficients and the "enhancement" factors. This being so, the relative magnitude of the various mechanisms of flow can be evaluated. Lemon (1956) pointed out that any method of reducing evaporative losses of water from soil must somehow alter the mechanisms of water movement in the surface zone. A better understanding of these mechanisms and their magnitudes may indicate methods for their alteration and lead to a method of reducing evaporative losses of water from soil.

ACKNOWLEDGMENTS

The assistance of R. J. Reginato, B. A. Kimball, F. S. Nakayama, and H. L. Mastin in conducting these experiments is gratefully acknowledged.

LITERATURE CITED

Brust, K. J., C. H. M. van Bavel, and G. B. Stirk. 1968. Hydraulic properties of a clay loam soil and the field measurement of water uptake by roots: III. Comparison of field and laboratory data on retention and of measured and calculated conductivities. Soil Sci. Soc. Amer. Proc. 32:322–326.

Buckingham, E. R. 1907. Studies on the movement of soil moisture. USDA Bureau of Soils, Bull. 38.

Cary, J. W. 1966. Soil moisture transport due to thermal gradients: Practical aspects. Soil Sci. Soc. Amer. Proc. 30:428–433.

Hillel, D. I. 1968. Soil water evaporation and means of minimizing it. Final technical report submitted to the U. S. Dep. of Agr. under Project no. AID-SWC-32.

Jackson, R. D. 1964a. Water vapor diffusion in relatively dry soil: II. Desorption experiments. Soil Sci. Soc. Amer. Proc. 28:464–466.

Jackson, R. D. 1964b. Water vapor diffusion in relatively dry soil: III. Steady-state experiments. Soil Sci. Soc. Amer. Proc. 28:467–470.

Jackson, R. D. 1965. Water vapor diffusion in relatively dry soil: IV. Temperature and pressure effects on sorption diffusion coefficients. Soil Sci. Soc. Amer. Proc. 29:144–148.

Jackson, R. D., R. J. Reginato, and C. H. M. van Bavel. 1965. Comparison of measured and calculated hydraulic conductivities of unsaturated soils. Water Resour. Res. 1:375–380.

Kimball, B. A., and R. D. Jackson. 1971. Seasonal effects on soil drying after irrigation. Hydrology and water resources of Arizona and the Southwest. Proc. of the Ariz. Sect., Amer. Water Res. Ass. and the Hydrology Sect., Ariz. Acad. of Sci. 1:85–98.

Kimball, B. A., and E. R. Lemon. 1971. Air turbulence effects upon soil gas exchange. Soil Sci. Soc. Amer. Proc. 35:16–21.

Klute, A. 1973. Soil water flow theory and its application in field situations. Chap. 2 (this book) *In* R. R. Bruce et al. (ed.) Field soil water regime. SSSA Spec. Publ. no. 5. Soil Sci. Soc. Amer., Madison, Wis.

Lemon, E. R. 1956. The potentialities for decreasing soil moisture evaporation loss. Soil Sci. Soc. Amer. Proc. 20:120–125.

Millington, R. J., and J. P. Quirk. 1959. Permeability of porous media. Nature 183:387–388.

Philip, J. R. 1957. Evaporation, and moisture and heat fields in the soil. J. Meteorol. 14:354–366.

Philip, J. R., and D. A. de Vries. 1957. Moisture movement in porous materials under temperature gradients. Trans. Amer. Geophys. Union 38:222–232.

Rose, C. W. 1968a. Water transport in soil with a daily temperature wave. I. Theory and experiment. Aust. J. Soil Res. 6:31–44.

Rose, C. W. 1968b. Water transport in soil with a daily temperature wave. II. Analysis. Aust. J. Soil Res. 6:45–57.

Rose, C. W. 1968c. Evaporation from bare soil under high radiation conditions. Int. Congr. Soil Sci., Trans. 9th (Adelaide, Aust.) I:57–66.

van Bavel, C. H. M., L. J. Fritschen, and R. J. Reginato. 1963. Surface energy balance in arid lands agriculture, USDA Production Res. Report 76. Washington, D. C.

Wendt, C. W., T. C. Olson, H. J. Haas, and W. O. Willis. 1970. Soil water evaporation. *In* Evapotranspiration in the Great Plains. Research Committee Great Plains Agr. Counc. Publ. 50. Kansas State Univ., Manhattan, Kansas.

Experiments in Predicting Evapotranspiration by Simulation With a Soil-Plant-Atmosphere Model (SPAM)[1]

4

E. R. LEMON, D. W. STEWART, R. W. SHAWCROFT
AND S. E. JENSEN[2]

ABSTRACT

From extensive field study, we have introduced a comprehensive mathematical model that acts like a plant community. It is based upon the conservation of energy. Our understanding and deficiencies have been gauged by testing model forecasts of local climate and community processes against real world experience with a simple system—a corn field (*Zea mays* L.). Microclimate prediction is biologically good enough, but reveals inadequacies of understanding airflow fluid dynamics within the vegetation stand. The inability to measure or predict the degree of wetness of the soil surface hampers correct forecast of evaporation. Probably the most difficult problem to resolve is the biological one of predicting how leaf pores (stomates) open and shut under drouth stress, thus affecting both evaporation and photosynthesis in leaves. Additional serious problems will arise in the future modeling of nonuniform or more complex systems especially in forecasting the distribution of wind, momentum, and radiation within the foliage stand.

INTRODUCTION

A little over 10 years ago, the U. S. Department of Agriculture joined with Cornell University in extensive field studies to understand how plant communities interact with the environment. It was a team effort to study first the component parts of physical and physiological processes under field conditions. Once there was sufficient knowledge and expertise, the team was able to study simultaneously the component parts all together. A suitable computer model was developed to simulate the plant community—both its

[1]Contribution of the Northeastern Region, Agricultural Research Service, U. S. Department of Agriculture in cooperation with the New York State Agricultural Experiment Station at Cornell University, Ithaca, N. Y. Agronomy Department Series Paper no. 970.
[2]Research Soil Scientist, USDA and Professor, Cornell University, Ithaca, N. Y.; Research Scientist, Canada Department of Agriculture, Swift Current, Saskatchewan; Research Soil Scientist, USDA, Akron, Colorado; and Climatologist, Royal Veterinary and Agricultural University, Copenhagen, respectively.

environment and its interaction—then it was tested against the integrated field measurements. The work was done in a corn field (*Zea mays* L.) in Ellis Hollow, New York, 8 km (5 miles) east of the University. We report here the status of progress and unresolved problems.

Primarily, we have aimed at answering agricultural problems of water conservation and crop production. We wished to know, for example, what plant shape or crop architecture would best use a given climate for net photosynthesis and efficient water use. While such questions are no less important today in a world of expanding population and increased demands for fresh water, other significant applications have developed from society's growing awareness of environment-related problems.

Our work has had meteorological application because solar energy exchange at land surfaces is under direct control of plant and soil characteristics. Evidently knowledge of how solar radiation, absorbed at the earth's surface, is parcelled to heat air and evaporate water is needed to understand not only large-scale meteorological processes but local climate formation. Since evaporation of water plays such an important role in the hydrologic cycle, forecast of its "use" is also of interest to foresters, hydrologists, irrigationists, and water resource planners.

Our study of carbon dioxide exchange of growing crops is of geophysical interest. Calculations indicate that green plants growing on the land dampen present increases in atmospheric carbon dioxide due to fossil fuel burning. Now we recognize, in addition, how local plant communities "air condition" the air, removing noxious contaminants and adjusting temperature and humidity.

It makes sense to treat crops as energy exchange systems because the major plant processes are also solar energy driven. Photosynthesis uses sunlight energy to fix carbon dioxide into carbohydrate materials. When carbon dioxide and oxygen gases are transferred across wet interfaces to an aerial environment in both photosynthesis and respiration, water is unavoidably lost in energy use by a water transfer process called transpiration. In fact, transpiration accounts for large shares of the water lost from the land and the energy transformed from absorbed solar radiation when water is plentiful. If one makes an energy balance based on the energy conservation law, the portions of net absorbed solar radiation going to various energy forms for summertime eastern United States might be: 1 to 5% to photosynthesis; 40 to 90% to evaporation; 10 to 60% to heat air; and 5 to 10% to minor storage terms. How these forms are proportioned largely rests on water supply.

Many workers have quantitatively predicted individual components of the energy balance. Evaporation formulae are examples. Variants of one developed by H. L. Penman in 1948 are probably the most sound physically and most popular today (19, 36, 51, 56). Net photosynthesis models based only on absorbed sunlight are numerous (8, 11, 30, 34, 41). Recent net photosynthesis models adding aerodynamic carbon dioxide terms have appeared (10, 22, 31, 32, 54). Nevertheless, prediction modeling of all the surface energy exchange processes has proven, until now, too complex a problem to define quantitatively in terms that are reasonably sound, physically and physiologically. Interaction has been the chief difficulty in the modeling machinery.

Within the past 2 years, several comprehensive plant community-environmental interaction models based upon the energy balance have appeared in the literature (18, 27, 33, 42, 46, 47, 58). All of them take advantage of computer simulation techniques where a tremendous number of interaction calculations and iterations are made. Only one of them has had sufficient field testing to pinpoint areas of needed research for future model evolution. We look at this model and its testing next.

THE SOIL-PLANT-ATMOSPHERE MODEL (SPAM)

The computer simulation model SPAM was developed by Stewart (25, 43, 44, 45, 46, 47, 48) while its testing involved a team of several individuals.

Figure 1 gives the essential ingredients of the model and its predictions. Let us take up the latter first to stress what the model can and cannot do. SPAM can answer questions in two general areas: (i) it can forecast the microclimate in a community and at soil and leaf surfaces with various leaf and community traits and external climates, and (ii) it can predict activity of leaves or of the community, such as, respiration, photosynthesis, evaporation and transpiration, heat dissipation, and noxious gaseous absorption.

Some predicted climate properties, which can change vertically through a plant community during midday, are pictured in Fig. 1 as "profiles" in the lower portion of the prediction box. One sees from left to right profiles of wind (u), light (Lt), carbon dioxide (C), water vapor (e), and air temperature (T). These are forecasts of steady-state mean values on a time scale of 1 hour. They define living conditions for plants and other organisms at any given level in the community, such as, walking animals, flying bugs, or "creeping" fungi living on a leaf.

In predicting activity, SPAM gives community processes in terms of source and sink intensities at any horizontal plane or vertical flux or flow densities across any horizontal plane. They can be defined for mass (i.e., water vapor and carbon dioxide), energy (radiation, latent and sensible heat, and photochemical energy equivalent), or momentum (wind shear). We picture activity of carbon dioxide (CO_2) and water vapor (H_2O) in the top portion of the prediction box. On the left, one sees that CO_2 flows both up and down in the midday case. Carbon dioxide diffuses upward from respiration in the soil and from poorly lit bottom leaves. It diffuses downward from the atmosphere to well-lit photosynthesizing upper leaves of the canopy. By convention, flux upward is plus and downward is minus. For water vapor, daytime upward flow steadily increases from the soil surface through to the top of the plant stand. Flux densities are often in mass or energy units per time per ground area.

Source and sink activity of CO_2 and H_2O are shown as source plus and sink minus. Thus, soil and plant respiration gives off CO_2 in the base of the stand as source and photosynthesis in the upper canopy is the CO_2 sink. Water vapor from soil evaporation is the source at the base, and transpiration is the water vapor source in the canopy. Source and sink intensities are usually in mass or energy units per time per volume. Again, the quantities are for mean steady state on a time scale of 1 hour.

Forecast of evaporation has many areas of application. In agriculture, water conservation aims at more effectively using water in unavoidable evaporation and transpiration. SPAM can help design crops and cropping schemes to do this. It can improve irrigation planning and scheduling, too. However, we are learning from big cities paved with concrete that, in the broader sense, evaporation and transpiration are desirable for proper air conditioning. Thus, in agriculture, we do not want to stop evaporation, just obtain the most effective use from it.

In forecasting net photosynthesis, SPAM can help the plant breeder and agronomist select more efficient plant shapes and planting patterns. Questions can be answered about crop adaptation to new lands and new crops or new cropping sequence in established areas undergoing social and economic change. Farmer harvest schedules for forage crops can be optimized. With manipulation of the submodels within SPAM, questions about feasibility, desirability, and sensitivity of factor changes on final outcome can be tested. For example, you can determine whether changing the leaf angle of a crop has as much influence on net photosynthesis as changing the individual leaf photosynthesis response to light.

We need to stress that while SPAM can predict net photosynthesis or dry matter gain, it is not a plant growth or crop yield model. While net photosynthesis is a major building block to growth and yield, it is but one of several processes involved. C. T. de Wit (8) of the Netherlands is a leader in complex growth model development.

Ideally, SPAM should be able to answer questions of community environment and activity for systems of any size, shape, or external climate. It cannot. It gives reasonable answers only for systems that are (i) simple or uniform in structure, (ii) large enough in extent to avoid horizontal climate variation, and (iii) under steady-state or slowly changing conditions. Extensive dense and vigorous agricultural crops approach conditions (i) and (ii). Clear or cloudy days approach condition (iii) except near sunrise and sunset.

Boundary Conditions

Now we turn to inputs of SPAM, then see how it operates. SPAM has two boundaries like a large horizontal slab with a top and bottom. The top is a plane in the airstream, 1 to 4 m above the stand. The bottom is the soil surface. At the top, external climate is defined by solar radiation, wind, temperature, carbon dioxide, and humidity. At the soil, we need to know heat storage, carbon dioxide evolution, and soil surface wetness, *SM*. From the

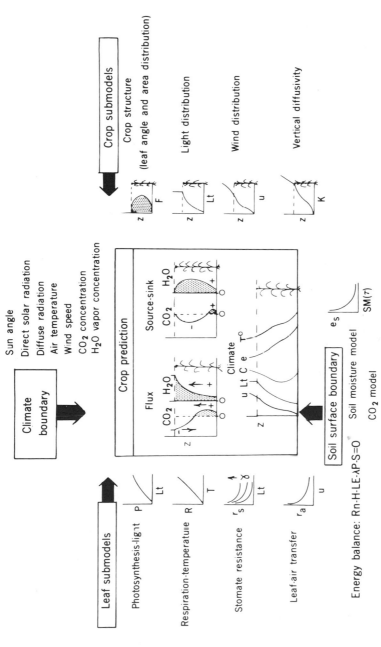

Figure 1. Schematic summary of a mathematical soil-plant-atmosphere model (SPAM) giving required inputs, submodels, and representative daytime predictions of climate and community activity (that is, water vapor and carbon dioxide exchange). Abbreviations: height (z), wind (u), light (Lt), concentration of carbon dioxide (C), water vapor (e), air temperature (T^o), surface vapor pressure (e_s), surface soil water potential $[SM\ (\tau)]$, photosynthesis (P), respiration (R), leaf temperature (T), stomate resistance (r_s), minimum stomate resistance at high light intensities (γ), gas diffusion resistance (r_a), leaf surface area (F), vertical diffusivity (K), net radiation (R_n), latent heat (LE), sensible heat (H), photochemical energy equivalent (λP), and soil heat storage (S).

latter, SPAM calculates in a submodel the apparent surface vapor pressure, e_s. Later, we see this item is more important to forecast evaporation from the soil and one of the most difficult to obtain.

The logic of SPAM is to (i) define on a leaf scale at many levels in the canopy, how each level, or the soil surface, will act in response to a given immediate climate; (ii) calculate from meteorology what that climate is; (iii) calculate the leaf and soil response to it, then; (iv) add up the leaf and soil responses, layer by layer for the whole stand. An energy balance is made on each layer, then on the stand as a whole by computer iteration. To do all this, certain information is needed both on the leaf scale and the stand scale.

Leaf Scale Submodels

Leaf scale submodels are on the left in Fig. 1. For photosynthesis response (P) of individual leaves to incident light (Lt), Chartier's model (6) has been modified to incorporate a stomate control mechanism. Carbon dioxide response is included in the photosynthesis submodel. For respiration response (R) to temperature (T), Waggoner's approach is used (57).

Kuiper's stomate opening relationship (20) is pictured in the leaf scale submodel defining the stomate resistance to gas diffusion (r_s) in response to light (L_t). Stomates are little valves on leaf surfaces controlling the passage of water vapor, carbon dioxide, and oxygen between wet inside membranes and dry outside air. With no water stress, stomates open in daylight and shut at night. However, as stress develops under water shortage, stomates close in daylight as a protective measure to water loss. Carbon dioxide diffusion is also cut. Shawcroft (43, 44) has modified Kuiper's model to include drouth effects, shown by a family of curves for increasing stress, gamma. This improvement is an empirical one for a very complex process. Lack of quantitative knowledge here is perhaps the weakest biological link in SPAM. The status of stomates rigidly controls transpiration and photosynthesis, and thereby, the whole energy balance. We cannot overstate this fact.

The final leaf scale submodel deals with the gas diffusion resistance (r_a) through the film of still air on the leaf surface. For this, Pohlhausen's formula, as derived by Gebhart (12) has been modified to account for natural field turbulence. Our experiments indicate a sizeable reduction in r_a with increasing windspeed (u) in turbulent air (46, 47, 49).

Crop Scale Submodels

Next, crop or community scale structure and submodels are pictured on the right in Fig. 1. The submodels deal with meteorological processes once the crop structure or architecture is defined.

The description of crop structure in quantitative terms is important and difficult (4, 28, 32, 34, 35). It is expressed in terms of plant surface area (F), distributed in height (z), and how it is displayed. Leaf size, leaf angle, and azimuth are all required. An adequate definition can only be made of simple stands where area distribution is uniform horizontally and reasonably uniform vertically. Nonuniformity, where clumping and vegetation gaps occur, creates special problems for light models predicting light distribution in the canopy.

To forecast light (Lt) with height in the stand (z), Duncan's model is used (11, 49). With modifications, infrared portions of the solar spectrum are predicted. Finally, thermal radiation is assessed from surface temperatures. The latter hinges on energy balance iteration. All three radiation regimes are needed to give net absorbed radiation for energy balance.

Because wind must diffuse gases and heat by turbulent motion, SPAM has to calculate windspeed distribution, u vs. z, and vertical turbulence diffusivity, K vs. z. Windspeed profiles above the stand first are generated by a method of Swinbank (50) knowing stand aerodynamic roughness traits as well as energy balance components. Profiles of wind in the stand are next predicted. This rests on distributing the wind drag on plant surfaces from the top of the stand downward into the canopy using a method of Perrier (37). Vertical diffusivity (K) is calculated by using a constant relation between it and the predicted wind in the stand as Cowan (7) has done.

Finally, the whole scheme must obey the energy balance in which energy sources must equal energy sinks. In this case (see Fig. 1), the net absorbed radiation is the driving source. The sinks are sensible heat, latent heat, photochemical energy equivalent, and soil heat storage (22). In the computer program, the energy balance is solved for each leaf layer and soil surface as well as for the entire system.

Solving the equation of any given part of SPAM is dependent on solving the equation of some other part. This interdependence requires the use of successive approximations in order to solve all of the equations simultaneously. The converging solutions thus give final answers for the complete system. Figure 2 gives the general procedure as a flow diagram.

THE TESTING OF SPAM

An experimental test of a mathematical model is an indispensable part of its proper development. We choose our best data to test SPAM for its weakness. On 18 August 1968, we took all the data needed in a 10-ha corn field in Ellis Hollow. 'Cornell M3' (*Zea mays* L.) is an ideal crop with leaves randomly oriented and relatively uniformly distributed in size and display. We planted it in a hexagonal array like an orchard so that all plants occupied equal space of 6 plants/m^2 land area. The crop was fully grown with a leaf area index (LAI) of 3.6 m^2 leaf area/m^2 land area. It was healthy but under mild water stress. The test day was prefectly clear with ideal growing season weather.

R. B. Musgrave took the needed leaf scale measurements of the photosynthesis submodel for us (17) while a team of several individuals made field

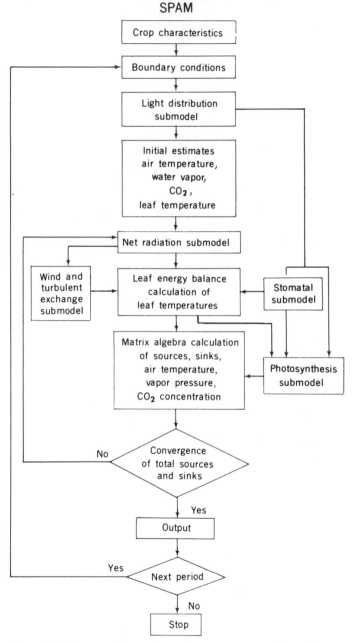

Figure 2. The general procedure of SPAM, given as a flow diagram.

measurements of climate profiles along with vertical fluxes to compare with model calculations. All of the SPAM requirements such as, structure, optical and fluid dynamic properties, and stomate response were measured on the

Figure 3. Equipment used to measure air temperature, water vapor, carbon dioxide concentration, and wind speed at various heights in and above uniform agricultural crops (soybeans shown here). Measurements are used to calculate items (photosynthesis, transpiration, and sensible heat exchange) in the energy balance, where the source is solar radiation.

same crop. Figure 3 shows some of the equipment used (in a field of soybeans) to measure the climate profiles in corn.

Theory vs. test is shown in Fig. 4, 5, and 6. Figure 4 compares forecast climate as solid and dotted lines against dashed lines drawn through data points. Size of error in data points is about equal to circle diameters. All data are 0.5-hour means spanning noon. Leaf area density, F, is given on the leaf and a scaled plant on the right for reference. All profiles have been adjusted to data points at the 240-cm height, judging this to be the best reference level. On this basis, SPAM undershoots temperature about 0.25C through most of the stand and overshoots water vapor about 0.5 g/m^3. It comes close to the mark on wind and CO_2 except near the soil.

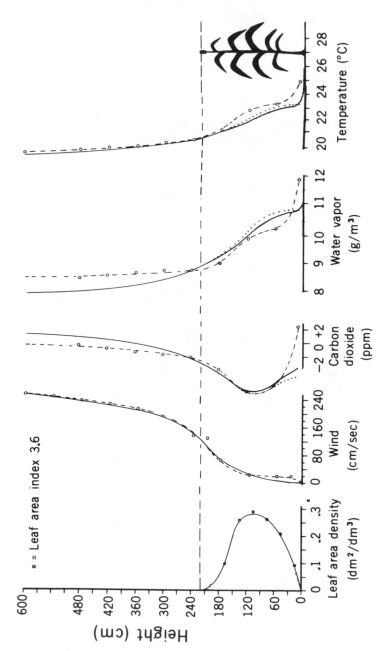

Figure 4. Measured (circles and dashed lines) and predicted (solid and dotted lines) profiles of climate factors in and above a corn field, with the field's vertical leaf area density constructed profile shown. Profiles are 0.5-hour mean values. (18 August 1968; 1145 to 1215 hours, Eastern Standard Time.)

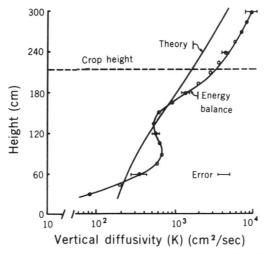

Figure 5. Measured energy balance profiles of vertical diffusivity (K) (circles and solid line) and model prediction (solid line) from momentum balance. (18 August 1968; 1145 to 1215 hours, Eastern Standard Time.)

The serious spread between predicted and measured water vapor and CO_2 above the stand raises questions about theory despite good wind and temperature profiles. Since we are confident of our data, we wonder about site and aerodynamic theory. Downwind fetch over the crop to sensors ran about 200 to 250 m, adequate for a boundary layer of 2 to 3 m over the stand. We question more seriously if classical fluid dynamic theory for boundary flow is applicable to tall vegetation that is porous and flexible. We suspect that troubles may start within the canopy since we cannot explain airflow deep in the stand. In Fig. 4, you can see in the real wind profile an almost constant wind from the densest part of the stand down to near the soil. This is indicative of no wind drag on the vegetation in the bottom half of the stand, which is physically impossible. Evidently some aerodynamic mechanism adds entrained air at the base of the stand for more airflow. To see if this explains other profile differences, we plug real wind into SPAM and reforecast the dotted profiles. Changes are slight. Next, we look to the vertical diffusivity (K) for answers. Vertical diffusivity is defined as an eddy diffusion coefficient. It is arrived at by assuming that transport of heat and water vapor can be expressed as (i) diffusion equations of time-averaged flux, gradients, and a coefficient (diffusivity) in the energy balance, or (ii) as time-averaged flux of momentum, wind gradients, and coefficients (diffusivity) in the momentum balance theory used in SPAM. The application of diffusion equations in both cases can be questioned if vertical mass flow occurs in the canopy. The additional entrained air in the base of the canopy suggests that mass flow does occur; now turbulent diffusivity values arrived at by either method will not account for the two transport mechanisms. Fortunately, from an operational point of view, our lack of understanding, and the incor-

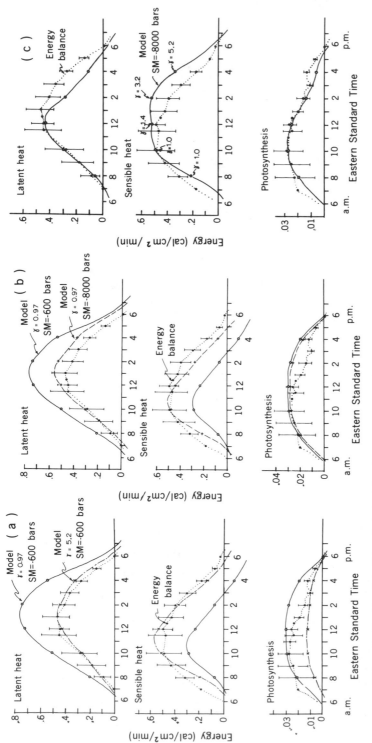

Figure 6. Predicted (solid and dashed lines) and measured (dotted lines) energy fluxes from a corn field during a clear summer day.. (*a*) Fixed constant soil surface water [$SM = -600$ bars (wet)] and two constant minimum stomatal resistances ($\gamma = 0.97$ and 5.2 sec/cm). (*b*) Two fixed constant soil surface water conditions [$SM = -600$ bars (wet) and -8000 bars (dry)] and constant minimum stomatal resistance ($\gamma = 0.97$ in sec/cm). (*c*) Fixed constant soil surface water [$SM = -8000$ bars (dry)] and measured minimum stomatal resistance (γ is measured in seconds per centimeter). (18 August 1968; length of bar denotes margin of error.)

rectness of the estimates of vertical diffusivity have little effect on outcome in a uniform crop. However, it is serious in a multistory forest (13, 24).

Figure 5 compares the vertical diffusivity predicted by SPAM and values computed from measurements of energy balance (14, 15, 16, 23, 53, 59). The latter's odd variation with height is dictated by the measured constant temperature and water vapor in midstand, as well as by the measured added airflow at the base. However, we have already shown that the additional mean airflow at the base does not appreciably alter the other predicted profiles. Thus, unexplained traits of fluid dynamics are at work both in the mid-canopy and at the bottom (1, 3, 9, 21, 38, 52, 55, 59, 60).

At the base of the canopy, errors in the temperature and water vapor profiles are traceable to a relatively small error in predicting net radiation and to a large error in predicting soil surface water potential. Although we have measured soil water beneath the surface, we have not been able to predict soil surface water in Ellis Hollow. Over 50% of the soil surface is occupied by flat stones which create an abnormally hot, dry surface in the daytime despite adequate and measurable water below the surface.

However, stones are only part of the problem. SPAM estimates vapor pressure at the soil surface from the apparent surface water potential of the soil, in a range where vapor pressure is substantially less than saturation. In this situation, the vapor flux is influenced by the dynamic time and space distributions of the heat and water fields below the surface. We have not as yet incorporated enough theory into SPAM to be able to deal with this complexity. Error in estimating CO_2 evolution at the base of the canopy has little effect on the outcome (2).

Despite inadequacies in theories of fluid dynamics and soil water, the prediction of the stand's climate is probably sufficient from a biological point of view.

Figure 6 gives corn crop activity on the ideal day, comparing SPAM output fluxes from the whole stand to energy balance fluxes from real data. In parts (a) and (b), we do a sensitivity test for wet and dry soil, as well as stress and no stress stomates, plugging into SPAM two fixed SM and two fixed gamma values for the whole day. Part (c) has real gamma values in SPAM but a fixed SM input. Comparing a wet case, $r_s = 0.97$ sec/cm, when the corn stomates are wide open, to a mild stress case where stomates are partially closed, $r_s = 5.2$ sec/cm, the latent heat flux is reduced for the day from 350 cal/cm^2 to 200 cal/cm^2 or 43%. One sees (in Table 1) that sensible heat increased from 57 units to 204 units, and net photosynthesis was reduced 37%. By drying the soil surface from a damp -600 bars to a dry -8000 bars, latent heat flux is reduced 38% and net photosynthesis 6%. Sensible heat flux increased from 57 units to 171 units.

It is obvious now (from Table 1) that both soil surface wetness and stomate status have rigid control over the sun's energy division into sensible and latent heat. Stomates also have rigid control over net photosynthesis, but apparent soil surface wetness has only a small effect here through temperature influence on respiration.

By putting the real stomata r_s values for the day in SPAM, prediction of

Table 1. A daytime energy balance for a corn crop, 18 August 1968, Ellis Hollow, New York. (Total incident solar radiation is 696 calories/cm^2; total net radiation, 453 calories/cm^2.)

Cases	Latent heat, cal/cm^2	Sensible heat, cal/cm^2	Photo energy, cal/cm^2	Soil storage, cal/cm^2
		SM(τ)* = -600 (Fig. 6a)		
Case 1				
γ[†] = 0.97	350 (0.77)[‡]	57 (0.12)	15.0 (0.033)	33 (0.074)
γ = 5.2	200 (0.45)	204 (0.45)	9.4 (0.021)	33 (0.074)
		γ = 0.97 (Fig. 6b)		
Case 2				
SM(τ) = -600	350 (0.77)	57 (0.12)	15.0 (0.033)	33 (0.074)
SM(τ) = -8,000	219 (0.50)	171 (0.39)	14.1 (0.032)	33 (0.075)
		SM(τ) = -8,000 (Fig. 6c)		
Case 3				
γ = measured values[§]	147 (0.33)	252 (0.57)	11.1 (0.025)	33 (0.074)
Measured energy balance	186 (0.41)	222 (0.49)	12.3 (0.027)	33 (0.073)
	±53	±53	±4.7	

* SM(τ) is surface soil water potential, measured in bars.
[†] γ is minimum stomatal resistance at high light intensities, measured in seconds per centimeter.
[‡] Figures in parentheses are fractions of net radiation.
[§] Measured γ are in Fig. 6c and 7.

net photosynthesis agrees with energy balance, giving us confidence that SPAM is working well. We are not as fortunate with latent and sensible heat since we cannot predict or measure soil water. By putting a fixed dry apparent surface soil water potential of −8000 bars into SPAM,[3] latent and sensible heat are close to the mark in the morning. In the afternoon, however, SPAM undershoots the mark indicative of a wetter effective surface or higher potential, −1200 bars. This makes sense because the effective vapor pressure will rise in the afternoon because of soil heating.

Despite the difficulties with soil water, the prediction of stomate status under stress, in the long run, is probably a far more difficult problem for biology. Figure 7 shows the complex trend of r_s through the moderate stress day. Under no stress, we know that r_s would follow the prediction curve. Since the leaves had undergone previous stress, they exhibited a sluggishness in early morning by not opening completely until 0.14 light units at 0800 despite ample water. Shortly after high light of 1200, the stomates gradually closed upon running low on water. Then they regained some water in the later afternoon and reopened somewhat at 0.08 light units at 1700 before sundown closure. These data are indicative of more than one complex feedback system. Several researchers (29, 40) have shown that light, water, temperature, and carbon dioxide are involved. They have reported on aging and hormonal control, as well (29, 40).

[3] The original model of SPAM had an error that made the surface soil water potential of −8000 bars reasonable for a dry soil. However, a revised version has corrected this to −2000 bars.

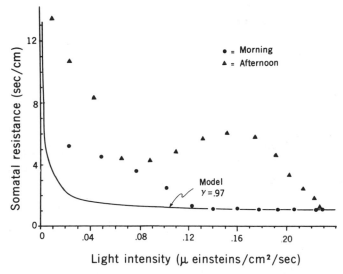

Figure 7. Stomate light response of corn leaves under mild water stress (data points). Resistance to gas diffusion through stomates is r_s (measured in seconds per centimeter). Solid line is prediction curve for the no-stress case, when corn stomates are wide open under bright sunlight [γ minimum stomatal resistance = 0.97]. Wet leaf $r_s = 0$ sec/cm (18 August 1968).

Can we believe SPAM's activity forecasts in Fig. 6 with the knowledge that all is not well with fluid dynamics and that errors can be hidden by our assuming soil surface wetness? Perhaps we can hazard a cautious "yes" based upon three arguments: (i) net photosynthesis comes out well while we know it is relatively insensitive to soil surface wetness; (ii) latent and sensible heat which are sensitive to soil surface wetness respond diurnally in a physically sound way to realistic *SM* inputs for a stoney soil; and (iii) the predictions are what one should expect from previous experimental experience (5, 26).

SIMULATION EXPERIMENTS OF CROP STRUCTURE AND CLIMATE EFFECTS ON EVAPOTRANSPIRATION

We have already demonstrated the sensitivity of SPAM to stomatal status and soil surface wetness and emphasized our present inability to accurately predict evaporation because of weakness in our submodels in these areas. Now we would like to demonstrate the usefulness of SPAM as an experimental tool to test the sensitivity of other aspects where the two weak submodels are given constant inputs, i.e., stomatal gamma values = 0.97 and soil surface water potential, *SM,* = −600 bars are held constant. (We would like to acknowledge many helpful suggestions of R. M. Peart during this phase of the study.)

We report two simulation experiments of a corn crop. The first compares total evapotranspiration (total latent heat) and crop transpiration as

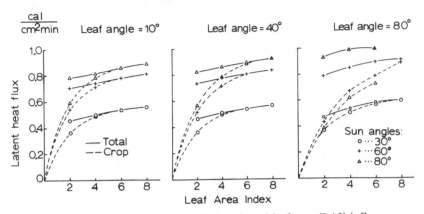

Figure 8. Simulation of corn crop leaf angle and leaf area (LAI) influence on evapotranspiration under various sun angles with other climatic conditions held constant. Climate is like that of Tampa, Florida or Corpus Christi, Texas, 28° N latitude. The broken lines indicate crop transpiration and the solid lines indicate crop transpiration plus soil evaporation.

affected by sun angle, leaf angle, and total leaf area (LAI). The second shows the influence of wind velocity, temperature, and relative humidity on total evapotranspiration and crop transpiration under constant noon-time radiation.

Figure 8 shows the results of the sun angle and crop structure interactions. Constant temperature (20.2C), relative humidity (66%), and wind speed (276 cm/sec) were assumed. Normal clear-day radiation distribution was used for a July day at 28° N, i.e., Corpus Christi, Texas and Tampa, Florida. All leaves at the three leaf angles were assumed to be uniformly and randomly distributed in space and azimuth. Also, all leaves in each angle class were assumed to be at the same angle (tilted up from the horizontal). The two assumptions about uniformity of leaf distribution in space and constancy of leaf angle are hardly realistic, especially for the higher leaf angle of 80° because row effect is not included in the model. Thus, caution must be used in interpretation, yet the analysis can be valuable if used with judgment. At the lower leaf angles, the model's assumption of random and uniform leaf placement is close to the real situation.

Figure 8 shows that the simulated total evapotranspiration increased only slightly with increasing leaf angle or increasing leaf area index (LAI > 2). In fact, the series of simulations showed an interchangeability between crop transpiration and soil evaporation. The decreased wind speed (not indicated) within the plant canopy caused by higher canopy densities should have very little effect on latent heat flux at 66% relative humidity. This conclusion can be reached from information in Fig. 9.

In Fig. 8, crop transpiration was extrapolated to a zero value at zero LAI since obviously no transpiration can occur without vegetation. Evapotranspiration was quite important at a low leaf area index (LAI = 2). As LAI increased, the canopy closed, and evaporation from the soil decreased. The

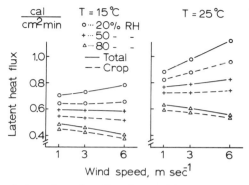

Figure 9. Simulation of varying climatic conditions on evapotranspiration of a corn crop with constant crop structure during midday radiation conditions in Ellis Hollow, New York. 18 August 1968. 42° N latitude.

figure also shows that crop transpiration actually decreased at high solar elevation angles (60° and 80°) when the leaves were erect (80°). More radiation could reach the soil under these conditions, and hence a much larger fraction of water loss came from the soil surface.

The data in Fig. 9 of the second simulation are perhaps of more significance. This arises because environmental variables are expectedly more important to evapotranspiration than is crop structure. In this simulation experiment, we used all the real corn crop data input we could for our 18 August 1968 test day at 1200 (see Fig. 4) but we varied the temperature, relative humidity, and windspeed at the upper reference boundary. As a point of interest, the 'Cornell M3' corn crop had a mean leaf angle slightly in excess of 40°. Solar radiation was 1.3 cal/cm^2 per min and net radiation was 0.96 cal/cm^2 per min. Soil surface wetness and stomatal status were held constant as in the previous simulation experiment.

In interpreting the Fig. 9 curves, one must ask whether certain combinations are likely to occur in nature, such as the high relative humidity, high wind velocity, low temperature, and clear weather. This also illustrates how easily unusual conditions can be simulated to check possibilities of environmental control in, for example, a greenhouse or with irrigation for relative humidity control. This could be especially valuable with a model accounting for water stress in plants. An interesting result shown is the different effect of increasing wind velocity at 25C and 80% relative humidity compared with 20% relative humidity. At the higher humidity, increasing wind decreased evapotranspiration, presumably by increasing the sensible heat loss from the leaves. Raschke discovered this phenomenon several years ago on single leaves (39).

All of the points shown in the last two figures total 51 computer runs and required only about 25 min of a large computer (IBM 360/65) time, a very inexpensive set of experiments. Studying such results can lead to better selection of variables for field experiments, improvement of the model itself, better understanding of the processes, and predictions of crop performance under new environments.

With caution, models are needed to solve complexity of stand structure. Their strength as a tool lies in evaluating the significance of isolated individual parameters under "controlled conditions" logistically impossible to achieve in the real world. Such information can help man order his priorities on selection traits.

ACKNOWLEDGMENT

We wish to recognize L. H. Allen, Jr. for his invaluable help in this research effort and his many suggestions in the preparation of the manuscript. Finally, we thank him for presenting this paper, substituting for the senior author, at the 1971 American Society of Agronomy symposium on Field Soil Moisture Regime, 16 August, in New York City.

LITERATURE CITED

1. Allen, L. H., Jr. 1968. Turbulence and wind spectra within a Japanese larch plantation. J. Appl. Meteorol. 7:73–78.
2. Allen, L. H., Jr., S. E. Jensen, and E. R. Lemon. 1971. Plant response to carbon dioxide enrichment under field conditions: A simulation. Science 173:256–258.
3. Allen, L. H., Jr., Edgar Lemon, and Ludwig Müller. 1972. Environment of a Costa Rican forest. Ecology 53:102–111.
4. Anderson, M. C. 1967. Photon flux, chlorophyll content, and photosynthesis under natural conditions. Ecology 48:1050–1053.
5. Brown, K., and W. Covey. 1966. The energy-budget evaluation of the micrometeorological transfer processes within a cornfield. Agr. Meteorol. 3:73–96.
6. Chartier, P. 1970. A model of CO_2 assimilation in the leaf. p. 307–315. *In* I. Setlik (ed.) Prediction and measurement of photosynthetic productivity. Pudoc, Wageningen, Netherlands.
7. Cowan, I. R. 1968. Mass, heat, and momentum exchange between stands of plants and their atmospheric enviornment. Quart. J. Roy. Meteorol. Soc. 94:523–544.
8. de Wit, C. T., R. Brouwer, and F. W. T. Penning de Vries. 1970. The simulation of photosynthetic systems. p. 47–70. *In* I. Setlik (ed.) Prediction and measurement of photosynthetic productivity. Pudoc, Wageningen, Netherlands.
9. Driclhet, A., and A. Perrier. 1971. Thermic action on transfer coefficients inside a canopy. Abstr.; Bull. Amer. Meteorol. Soc. 52:308–309.
10. Duncan, W. G., and B. J. Barfield. 1970. Predicting effects of CO_2 enrichment with simulation models and a digital computer. Trans. Amer. Soc. Agr. Eng. 13:246–248.
11. Duncan, W. G., R. S. Loomis, W. A. Williams, and R. Hanau. 1967. A model for simulating photosynthesis in plant communities. Hilgardia 38:181–205.
12. Gebhart, B. 1961. Heat transfer. McGraw Hill Book Co., New York. p. 454.
13. Geiger, R. 1965. The climate near the ground. 4th ed. Harvard Univ. Press, Cambridge, Mass.
14. Gillespie, T. J. 1971. Carbon dioxide profiles and apparent diffusivities in corn fields at night. Agr. Meteorol. 8:51–57.
15. Gillespie, T. J., and K. M. King. 1971. Night-time sink strengths and apparent diffusivities within a corn canopy. Agr. Meteorol. 8:59–67.
16. Groom, M., and E. R. Lemon. 1973. The energy balance model for calculating the gaseous exchange in crop canopies and estimates of errors. J. Appl. Meteorol. (in preparation).
17. Heichel, G. H., and R. B. Musgrave. 1969. Varietal differences in net photosynthesis of *Zea mays* L. Crop Sci. 9:483–486.

18. Idso, S. 1968. A holocoenotic analysis of environmental-plant relationships. Agr. Exp. Sta., Univ. of Minnesota, St. Paul. Tech. Bull. 264, p. 147.

19. Jensen, M. E., J. L. Wright, and B. J. Pratt. 1971. Estimating soil moisture depletion from climate, crop and soil data. Trans. Amer. Soc. Agr. Eng. 14:954-959.

20. Kuiper, P. J. C. 1961. The effect of environmental factors on the transpiration of leaves, with special reference to stomatal light response. Meded. Landbouwhogesch. Wageningen 61:1-49.

21. Landsberg, J. J., and A. S. Thom. 1971. Aerodynamic properties of a plant of complex structure. Quart. J. Roy. Meteorol. Soc. 97:565-570.

22. Lemon, E. R. 1967. Aerodynamic studies of CO_2 exchange between the atmosphere and the plant. p. 263-290. *In* Anthony San Pietro, Frances A. Geer, and Thomas J. Army (ed.) Harvesting the sun: Photosynthesis in plant life. Academic Press, New York.

23. Lemon, E. 1970. Mass and energy exchange between plant stands and environment. p. 199-205. *In* I. Setlik (ed.) Prediction and measurement of photosynthetic productivity. Pudoc, Wageningen, Netherlands.

24. Lemon, E., L. H. Allen, Jr., and L. Müller. 1970. Carbon dioxide exchange of a tropical rain forest, Part II. Bioscience 20:1054-1059.

25. Lemon, Edgar, D. W. Stewart, and R. W. Shawcroft. 1971. The sun's work in a cornfield. Science 174:371-378.

26. Lemon, E. R., and J. L. Wright. 1969. Photosynthesis under field conditions. XA. Assessing sources and sinks of carbon dioxide in a corn crop (*Zea mays* L.) using a momentum balance approach. Agron. J. 61:405-411.

27. Lommen, P. W., C. R. Schwintzer, C. S. Yocum, and D. M. Gates. 1971. A model describing photosynthesis in terms of gas diffusion and enzyme kinetics. Planta 98: 195-220.

28. Loomis, R. S., and W. A. Williams. 1969. Productivity and the morphology of crop stands: Patterns with leaves. p. 27-47. *In* J. Eastin, F. A. Haskins, C. Y. Sullivan, and C. H. M. van Bavel (ed.) Physiological aspects of crop yield. Amer. Soc. of Agron., Madison, Wis.

29. Meidner, H., and T. A. Mansfield. 1968. Physiology of stomata. McGraw-Hill, New York.

30. Monsi, M., and T. Saeki. 1953. Uber den Lichtfaktor in den Pflanzengesellschaften und seine Bedeutung fur die Stoffproduktion. Jap. J. Botany 14:22-52.

31. Monteith, J. L. 1962. Measurements and interpretation of carbon dioxide fluxes in the field. Neth. J. Agr. Sci. 10:334-346.

32. Monteith, J. L. 1969. Light interception and radiative exchange in crop stands. p. 89-111. *In* J. Eastin, F. A. Haskins, C. Y. Sullivan, and C. H. M. van Bavel (ed.) Physiological aspects of crop yield. Amer. Soc. of Agron., Madison, Wis.

33. Murphy, C. E., and K. R. Knoerr. 1970. Modeling the energy balance processes of natural ecosystems. Final Research Report 1969-1970. School of Forestry, Duke University, Durham, N. C. p. 164.

34. Nichiporovich, A. A. 1961. Properties of plant crops as an optical system. Soviet Plant Physiol. 8:536-546.

35. Niilisk, H., T. Nilson, and J. Ross. 1970. Radiation in plant canopies and its measurement. p. 165-177. *In* I. Setlik (ed.) Prediction and measurement of photosynthetic productivity. Pudoc, Wageningen, Netherlands.

36. Penman, H. L. 1948. Natural evaporation from open water, bare soil, and grass. Roy. Soc. London, Proc. Ser. A 193:120-146.

37. Perrier, E. 1967. Microturbulence et transferts dans les converts vegetaux. La Meteorologie I(4):527-550.

38. Perrier, E. R., R. J. Millington, D. B. Peters, R. J. Luxmoore. 1970. Wind structure above and within a soybean canopy. Agron. J. 58:615-620.

39. Raschke, K. 1956. Über die physikalischen Beziehungen zwischen Wärmeübergangszahl Strahlungsaustavsch, Temperatur und Transpiration eines Blattes. Planta 48: 200-238.

40. Raschke, K. 1970. Temperature dependence of CO_2 assimilation and stomatal aperture in leaf sections of *Zea mays*. Planta 91:336-363.

41. Ross, J., and T. Nilson. 1968. The calculation of photosynthetically active radiation in plant communities. p. 5-54. *In* Rezhim solnechnoy radiatsii v rastitelnom pokrov. IPA, Tartu, Estonia.

42. Seginer, Ido. 1971. Wind effect on the evaporation rate. J. Appl. Meteorol. 10: 215–220.

43. Shawcroft, R. W. 1970. Water relations and stomatal responses in a corn field. Ph.D. Thesis, Cornell University. 127 p. (Dissertation Abstracts order number 70–24,012, Univeristy Microfilms, Ann Arbor, Mich.)

44. Shawcroft, R. W. 1971. The energy budget at the earth's surface: Water relations and stomatal response in a corn field. U. S. Army ECOM Technical Report 2–68–1–7. U. S. Army Electronics Command, Fort Huachuca, Ariz. (Microfilm available from U. S. Department of Commerce, National Technical Information Center, Springfield, Va.). 94 p.

45. Shawcroft, R. W., and E. R. Lemon. 1972. Estimation of internal crop water status from meteorological and plant parameters. p. 449–459. *In* Plant response to climate factors. UNESCO Symp., Uppsala, Sweden. (Ecology and Conservation-5).

46. Stewart, D. W. 1970. A simulation of net photosynthesis of field corn. Ph.D. Thesis, Cornell University, Ithaca, New York. 132 p. (Dissertation Abstracts order number 70–17,101, University Microfilms, Ann Arbor, Mich.)

47. Stewart, D. W., and E. R. Lemon. 1969. The energy budget at the earth's surface: a simulation of net photosynthesis of field. U. S. Army ECOM Technical Report 2–68–1–6. U. S. Army Electronics Command, Fort Huachuca, Ariz. (Microfilm available from U. S. Department of Commerce. National Technical Information Center, Springfield, Va.). 132 p.

48. Stewart, D. W., and E. R. Lemon. 1973. A mathematical model of photosynthesis and energy exchange of homogeneous vegetation. J. Appl. Meteorol. (in preparation).

49. Stewart, D. W., R. W. Shawcroft, and E. R. Lemon. 1973. Effect of leaf angle and turbulence on the boundary layer resistance of corn (*Zea mays*) leaves. Agron. J. 66: (in press).

50. Swinbank, W. C. 1964. The exponential wind profile. Quart. J. Roy. Meteorol. Soc. 90:119–135.

51. Tanner, C. B., and M. Fuchs. 1968. Evaporation from unsaturated surfaces: A generalized combination method. J. Geophys. Res. 73:1299–1304.

52. Thom, A. S. 1971. Momentum absorption by vegetation. Quart. J. Roy. Meteorol. Soc. 97:414–428.

53. Uchijima, Z. 1962. Studies on the micro-climate within the plant communities. (1) on the turbulent transfer coefficient within plant layer. J. Agr. Meteorol. (Tokyo) 18:1–9.

54. Uchijima, Z., and E. Inoue. 1970. Studies of energy and gas exchange within crop canopies (9) Simulation of CO_2 environment within a canopy. J. Agr. Meteorol. (Japan) 26:5–18.

55. Uchijima, Z., and J. L. Wright. 1964. An experimental study of air flow in a corn plant-air layer. Bull. Nat. Inst. Agr. Sci. (Japan) Ser. A (No. 11)19:19–66.

56. van Bavel, C. H. M. 1966. Potential evapotranspiration: combination concept and experimental verification. Water Resour. Res. 2:455–467.

57. Waggoner, P. 1969. Environmental manipulation for higher yields. p. 343–373. *In* J. D. Eastin, F. A. Haskins, C. Y. Sullivan, and C. H. M. van Bavel (ed.) Physiological aspects of crop yield. Amer. Soc. of Agron., Madison, Wis.

58. Waggoner, P. E., G. M. Furnival, and W. E. Reifsynder. 1969. Simulation of the microclimate in a forest. Forest Sci. 15:37–45.

59. Wright, J. L., and K. W. Brown. 1967. Comparison of momentum and energy balance methods of computing vertical transfer within a crop. Agron. J. 59:427–432.

60. Wright, J. L., and E. R. Lemon. 1966. Photosynthesis under field conditions. VIII. An analysis of windspeed fluctuation data to evaluate turbulent exchange within a corn crop. Agron. J. 58:255–261.

Relationships Between Soil Structure Characteristics and Hydraulic Conductivity[1]

5

J. BOUMA AND J. L. ANDERSON[2]

ABSTRACT

Relationships between soil structure, as characterized by morphometric methods, and hydraulic conductivity (K) were explored for a series of six artificial sand-clay mixtures with characteristic basic fabrics and four pedal soil horizons. The hydraulic behavior was strongly affected by differences in basic fabrics. Pore size distributions (measured with a point-count technique in thin sections) could be used to predict K values of the soil mixtures, but a pore continuity model and large matching factors were used. A physical model, using moisture retention data, yielded comparable results with less effort. Morphometric analysis was shown to have a specific function in studying the occurrence of pore types, such as planar voids and channels, that constitute only a small fraction of total pore volume, but strongly affect K_{sat}. Results were used to construct simple models of natural soil structures. A planar-void model, assuming vertical plane continuity, was used to calculate K_{sat} of the four natural pedal soil materials. Results obtained were close to experimental values measured *in situ* by the double-tube method.

INTRODUCTION

Questions about the physical behavior of soil types and in particular their capacity to transmit water are becoming more specific as soil maps are increasingly used.

Field methods are available to measure hydraulic conductivity (K) of saturated and unsaturated soil well above the ground-water table (Bouwer, 1961, 1962; Bouwer & Rice, 1964, 1967; Boersma, 1965; Gardner, 1970; Bouma et al., 1971). Such methods are preferable to those requiring sampling of "undisturbed" cores in the field and subsequent measurement in the laboratory (Klute, 1965). The field techniques require instrumentation and are laborious. Hydraulic field characterization of major horizons is too costly to be applied to all soils represented on a soil map. An alternative could be to study horizons of benchmark soils and to extrapolate measured values

[1]Contribution from the Geological and Natural History Survey, University Extension, Madison, in cooperation with the Department of Soil Science, College of Agricultural and Life Sciences, University of Wisconsin, Madison, 53706. This research was supported in part by a grant from the Wisconsin Department of Natural Resources. Grateful acknowledgement is made to F. D. Hole for helpful suggestions.
[2]Associate Professor of Soil Science, and Research Assistant, respectively.

to other soils that are somehow characterized as being comparable. Such procedures would require a reasonable understanding of flow patterns of water through pores of soils. A detailed functional characterization of pore systems in soils is needed for these procedures. One method is to estimate pore size distributions of nonswelling soil materials from moisture retention data and to use this derived pore-size distribution in a pore model to predict K values (Marshall, 1958). This method has also been applied to swelling soil materials by Green and Corey (1971) who reported reasonable agreement between calculated and measured K values, both for saturated and unsaturated soil. Because calculated and measured values did not agree directly, it was necessary to measure one K value (K_{sat}) experimentally to determine a "matching factor". This is the ratio between the measured and calculated value. All other calculated values applying to unsaturated soil had to be multiplied by this factor.

However, many pores in soil can be observed directly and their sizes measured by morphological techniques. Different types of pores with supposedly different functions in the soil can be distinguished (Brewer, 1964). In addition, physical theory describes relationships between pore size and conductivity for planar and tubular pores (Childs, 1969). The purposes of this exploratory paper are (i) to investigate the feasibility of using morphometric data for predicting K values of soils and (ii) to review and refine methods for describing soil structure in the field and in the laboratory that will enable the investigator to make estimates of K from morphometric data.

MORPHOLOGICAL STUDY OF SOIL STRUCTURE

Definitions

The widely used field concept of soil structure (Soil Survey Staff, 1951) describes structure as "the aggregation of primary soil particles, into compound particles or clusters of primary particles, which are separated from adjoining aggregates by surfaces of weakness". Accordingly, soil materials without such compound particles are considered to have no structure and are either "massive" or "single grain". Micromorphological research has demonstrated that these terms are inadequate to describe unaggregated soil materials and their porosity (Kubiena, 1938; Jongerius, 1957; Brewer, 1964). Some types of voids, such as tubular pores, do not result from random packing of primary or compound particles and need to be considered as separate entities. Therefore, soil structure has been defined subsequently (Brewer, 1964) as

> the physical constitution of a soil material as expressed by size, shape and arrangement of the solid particles and the voids. Particles include both the primary particles that form compound particles and the compound particles themselves. Fabric is the element of structure that deals with arrangement.

Following this concept, voids in any soil material are a part of soil structure whether occurring between sand grains in an unaggregated loose sand or in

Table 1. Levels of description of soil pores, modified* after Brewer (1964).

Structure level	Unit of organization	Components* (excluding pedological features)
A. Plasmic structure	S-matrix or primary ped	Plasma
B. *Basic structure	S-matrix or primary ped	Plasma + skeleton grains + associated voids (simple packing voids)
C. *Matric structure	S-matrix or primary ped	Plasma + skeleton grains + all voids
D. Primary structure	S-matrix or primary ped	Plasma + skeleton grains + all voids + intra-pedal pedological features
E. Secondary structure	Secondary ped	Primary peds as units + interpedal voids
F. Tertiary structure	Tertiary ped	Secondary peds as units + interpedal voids

* Brewer's term: "basic structure" = matric structure of this table.

and between well-developed compound particles (peds) in a clay. Peds are defined by Brewer (1964) as

> individual natural soil aggregates consisting of a cluster of primary particles and separated from adjoining peds by surfaces of weakness which are recognizable as natural voids or by the occurrence of cutans. Primary peds are the simplest peds in a soil material; they cannot be divided into smaller peds but they may be packed together to form compound (secondary, tertiary) peds.

The concept of structure (Soil Survey Staff, 1951) is comparable to Brewer's pedality, which is "the physical constitution of a soil material as expressed by size, shape and arrangement of peds." Description of soil materials, according to Brewer's system, follows a logical sequence from small to large units (Table 1). As this study is concerned with relationships between soil structure and hydraulic conductivity, attention is confined to occurrence and morphology of soil pores; pedological features other than ped cutans are not discussed. The original definition of basic structure (Brewer, 1964) is modified to read: the physical constitution of a soil material as expressed by size, shape, and arrangement of the simple grains (plasma and skeleton grains) and the simple packing voids in primary peds or apedal soil materials (not occurring in pedological features other than plasma separations). The porosity in such basic fabrics is comparable to the "textural porosity" as referred to in Childs (1969) and is specific for any soil material (Bouma, 1969). Occurrence of vughs, channels, or other voids considerably larger than these packing voids is determined by variable external factors of structure formation. These larger pores are part of the proposed "matric structure" (Table 1) which is the structure of the *s*-matrix (Brewer, 1964) and includes all voids in primary peds or apedal soil material.

Applications

In this investigation, relationships between soil structure and hydraulic conductivity were studied at the levels of basic and secondary structures. Basic structures were observed in thin sections and secondary structures in soil peels prepared from undisturbed horizontal sections through selected soil horizons. Description of basic structures was made according to their related distribution pattern (Table 5) (Kubiena, 1938; Brewer, 1964). Results were quantified by using a point-count technique (see p. 81–86, this chapter). Secondary structures were described in the field by noting sizes, shapes, and arrangement of peds and abundance of channels (Table 3). Sizes and shapes of peds were classified according to the Soil Survey Staff (1951) and Brewer (1964). Arrangement of peds was described by noting (i) accommodation (a measure of the degree to which adjacent faces are molds of each other), (ii) packing (whether with some orderly arrangement or at random), and (iii) inclination (orientation with reference to the vertical or horizontal) (Brewer, 1964, p. 345).

Description of grade of structure (Table 3) (Soil Survey Staff, 1951) is difficult. The (i) "degree of evidence in place" and (ii) the durability of peds after an undefined procedure of "disturbance" result from processes of adhesion and cohesion which are mainly functions of variable moisture content. Ped durability may be expressed in terms of soil consistence (Soil Survey Staff, 1951). "Degree of evidence" of peds in place may be based on observation of arrangement of peds including accommodation and packing.

Descriptive morphological classifications of secondary structures are becoming very complex with increasing numbers of categories of size, shape, and arrangement. The more detailed such vocabulary becomes, the more terms are needed to describe natural variability in a horizon. This development may be unfavorable for applied studies insofar as one of the purposes of structure descriptions is to distinguish basic order in the overwhelming complexity of a pedal soil material. This points to the need for quantitative measurements of specific well-defined structural features. Abundance and sizes of channels were measured on soil peels by Van der Plas and Slager (1964) and Bouma and Hole (1965). Counts of large channels (> 1 mm diam.) on horizontal sections through soil horizons were made in the field by Slager (1966) and Baxter and Hole (1967). A technique is explored in this paper to quantify patterns of natural planar voids in soil peels (see p. 90 , this chapter).

In this study of relationships between void patterns and hydraulic conductivity, the attempt has been made to represent natural structures by simplified morphological models that use a few pore types (channels, planes, and packing voids) for which physical expressions are available to relate pore size to conductivity. In such models much emphasis has to be given to continuity of pores in the soil as this determines their effectiveness in transmitting water. Studies of pores in thin sections and in soil peels yield data that applies two-dimensionally. Statistical procedures are necessary to use this data in predicting the behavior of a three-dimensional volume of soil.

A simplified structural model in *apedal* soil materials (Fig. 1) has a basic structure in which packing pores are determined by size, shape, and arrangement of the plasma and skeleton grains. Larger pores, such as vughs, channels, chambers, and possibly some planar voids (Brewer, 1964), may occur in this basic structure in varying patterns (matric structure) and may contribute

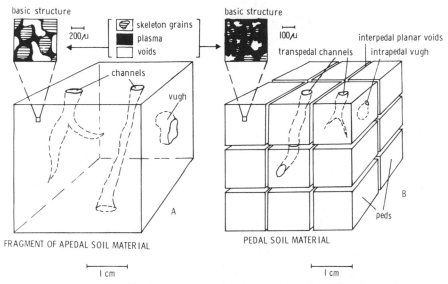

Figure 1. Simplified structure models in apedal and pedal soil materials.

to K depending on their size and continuity. A series of artificially prepared basic structures was investigated with morphological methods, and K values were measured in the laboratory to explore mutual relationships and to obtain equations for calculating K from morphometric data (Fig. 2). Models of *pedal* soil materials (Fig. 1) have an intrapedal porosity determined by the basic and matric structure of the soil material and an interpedal porosity which is commonly planar. Transpedal channels may penetrate several peds. Soil peels from four pedal soil materials were investigated with morphological methods, and K values were measured *in situ* in the field to obtain equations for calculating K_{sat} from morphometric data (Fig. 3).

SOIL MATERIALS AND PROCEDURES

Basic Structures

A series of soil materials was prepared by mixing medium sand with increasing quantities (10, 20, 30, and 40% by weight) of soil material from a B3 horizon of an Ontonagon silty clay loam. The textural composition of the mixtures and other physical properties are presented in Table 2. Water was added slowly during the stirring and kneading of each mixture of sand and clay. After reaching moisture contents corresponding to the lower plastic limit, the materials were kneaded thoroughly for a period of at least 0.5 hour to obtain complete mixing. Air-dried samples from each puddled mixture were used to prepare thin sections (Buol & Fadness, 1961). Large cylindrical columns were prepared from each material by shaping the plastic soil mass into cylindrical forms, 40 cm high and 7.5 cm in diameter. These columns,

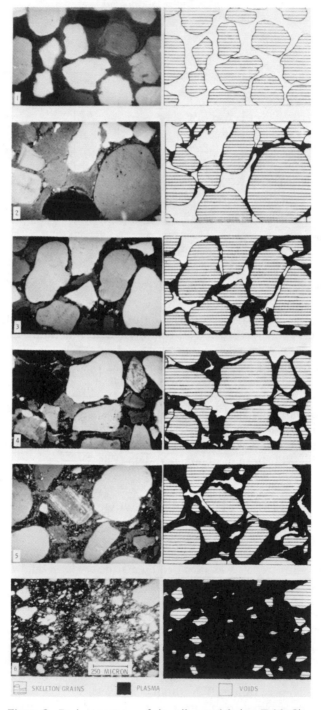

Figure 2. Basic structures of six soil materials (see Table 2).

Table 2. Textures and physical properties of soil mixtures used for studying K values of basic fabrics (Fig. 2).

No.	Clay	Silt f	m	c	Sand vf	f	m	c	vc	Textural class	Particle density	pH	Organic matter
					%								%
1	--	--	--	--	--	--	100	--	--	Sand	2.66	5.9	0
2	4.6	1.2	1.2	0.4	1.3	0.6	90.5	0.1	--	Sand	2.67		0.1
3	9.3	2.5	2.4	0.8	2.6	1.1	80.9	0.3	--	Loamy sand	2.68		0.2
4	13.8	3.6	3.6	1.2	4.0	1.8	71.4	0.6	--	Sandy loam	2.69		0.3
5	10.6	5.0	4.8	1.6	5.2	2.2	61.8	0.6	--	Sandy clay loam	2.70		0.4
6	46.5	12.5	12.0	4.0	13.0	5.8	4.6	1.4	0.2	Clay	2.75	7.5	0.8

to be used for the crust test (see p. 81–86, this chapter), were then slowly air dried. This procedure (Bouma, 1969) avoids many of the problems associated with the packing of artificial aggregates of arbitrary size into cylinders. The columns have a homogeneous structure throughout with only fine pores, the size of which is determined by the packing of the primary soil particles [textural porosity, Childs (1969) or basic fabric (see p. 78–81, this chapter)]. Large fragments of air-dried mixtures (about 200 to 300 cm^3) were slowly remoistened and used to determine moisture retention characteristics. Cylindrical cores, 5 cm high with a diameter of 7.5 cm, were prepared from the puddled mixtures in cylinders, removed, air-dried, slowly remoistened, and replaced in the cylinders for measurement of K_{sat}.

Secondary Structures

Soil peels (Bouma & Hole, 1965; Jager & van der Voort, 1965) were prepared from horizontal sections through selected soil horizons having well-developed pedality. A block of soil was carefully carved out *in situ* from the level of the section downwards to fit into a rectangular metal container. This was gently pushed down over the block of soil which was then cut loose from below and removed with the container. The soil surface was smoothed and a peel was prepared from it and mounted on masonite. General characteristics of the four studied horizons are reported in Tables 3 and 4.

MEASUREMENT AND CALCULATION OF K

Physical Methods

The crust test for measuring hydraulic conductivity of unsaturated soil materials (Hillel & Gardner, 1970) was applied in the laboratory using large columns composed of air-dried, previously puddled, and mixed soil materials (see p. 81–83, this chapter) with different successive crusts on top of them. The columns were placed on a ceramic porous plate to which different tensions could be applied. This proved necessary when measurements were made in the more permeable basic fabrics through crusts of low resistance (with corresponding low tensions in the soil below the crusts). If not re-

Table 3. Morphological characteristics of soil horizons sampled for studying K values of secondary structures (Fig. 3).

No.	Soil classification	Field structure		
		Soil survey manual	Brewer	Channels (counted in 625 cm^2 of exposed horizontal planes)
1 & 2	Typic Argiudoll (fine silty, mixed, mesic) Tama silt loam B3 (silty clay loam)	(strong) medium prismatic parting to (moderate) medium subangular blocky	Accommodated, faulted, medium tetracolumnar secondary peds, composed of accommodated, normal, medium, subrounded octablocky primary peds	6 fine (1–2 mm diam) 1 medium (2–4 mm diam)
3	Typic Argiudoll (fine silty, mixed, mesic) Plano silt loam B2 (silty clay loam)	(weak) medium prismatic parting to (moderate) fine subangular blocky	Accommodated, faulted, medium tetracolumnar secondary peds, composed of accommodated, normal, fine subrounded pentablocky primary peds	12 fine 10 medium 2 large (larger than 4 mm diam)
4	Same as 3 B3 (silty clay loam)	(moderate) coarse prismatic	Accommodated, bi-offset, medium pentacolumnar primary peds	4 fine 4 medium 1 large
5	Typic Eutrochrept (very fine, mixed, mesic) Oshkosh clay B3 (silty clay)	(moderate) medium prismatic parting to strong medium angular blocky	Accommodated, bi-offset, locally offset faulted, medium tetracolumnar secondary peds, composed of accommodated, normal, medium hexablocky primary peds	4 fine 1 medium
6	Typic Udipsamment (sandy, mixed, mesic) Plainfield loamy sand B2 (loamy fine sand)	Single grain	Apedal; basic fabric with intertextic related distribution pattern	4 fine

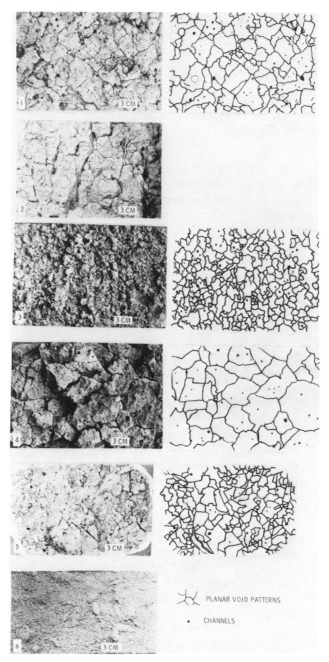

Figure 3. Secondary structures of four pedal soil materials. The numbers on the pictures correspond with those in Tables 3 and 4. Picture 2 gives soil structure in a vertical soil peel of soil no. 1. Picture 6 is included as an example of an apedal soil material at a comparable magnification (B2 of Plainfield loamy sand).

Table 4. Physical properties of soil horizons used for studying K values of secondary structures (Fig. 3).

No.	Type of soil	Horizon	Bulk density	Particle density	Porosity	Organic matter	pH
			1/3b	g/cm³	—— % ——		
1 + 2	Tama silt loam	B3	1.35	2.65	49.0	0.5	5.5
3	Plano silt loam	B2	1.41	2.63	46.3	0.4	5.2
4	Plano silt loam	B3	1.45	2.64	45.1	0.7	5.1
5	Oshkosh clay	B2	1.43	2.62	45.5	0.8	7.7
6	Plainfield loamy sand	B2	1.45	2.65	45.3	1.1	5.8

moved through the porous plate, water would have accumulated at the bottom of the column until saturation permitted out flow. Attendant capillary rise of water into the column would have invalidated the relationship between infiltration rate through the crust and subcrust tension.

K_{sat} was measured for the soil mixtures separately on large cores (see p. 81–83, this chapter) (Klute, 1965). The modified crust test was applied in the field to measure K of unsaturated soil *in situ* (Bouma et al., 1971). Moisture retention curves were determined with standard procedures on large Saran-coated fragments or field clods (Brasher et al., 1966). COLE values (Grossman et al., 1968), bulk density, and porosity were calculated from these data, the latter after determining particle density (Blake, 1965). Hydraulic conductivity was measured *in situ* in the field with the double-tube method (Bouwer, 1961, 1962; Bouwer & Rice, 1964, 1967; Baumgart, 1967) for measuring K_{sat} in soil well above the ground water. Hydraulic conductivity of saturated and unsaturated soil was estimated from moisture retention data using equation [1] from Green and Corey (1971) that is similar to the one given by Kunze, Uehara, and Graham (1968):

$$K(\theta)_i = (K_s/K_{sc}) \cdot (30\delta^2/\rho g \eta) \cdot (e^p/n^2) \cdot \sum_{j=1}^{m} \left[(2j + 1 - 2i)h_j^{-2} \right]$$

$$(i = 1, 2, \ldots .m)$$

[1]

where
$K(\theta)_i$ = the calculated conductivity for a specified water content (cm/min)
θ = water content (cm³/cm³)
i = last water content class on the wet end: $i = 1$ = pore class corresponding with θ_{sat}, $i = m$ = pore class with lowest water content for which K is calculated
K_s/K_{sc} = matching factor (= measured/calculated K)
δ = surface tension of water (dynes/cm)
ρ = density of water (g/cm³)
g = gravitational constant (cm/sec²)
η = viscosity of water (g/cm sec)
e = porosity (cm³/cm³)
p = parameter.

Here $p = 2$ (Marshall, 1958); $n =$ total number of pore classes between $\theta = 0$ and θ_{sat}; and $h_j =$ pressure of a given class of waterfilled pores (cm water).

Morphological Methods

BASIC STRUCTURES

A microscopic point-count technique was applied to obtain estimates of total volumes and size distributions of visible pores in a thin section (Chayes, 1956; Anderson & Binnie, 1961). One thousand two hundred points were counted per section in 240 fields with each five points at a linear magnification of 80X. Resulting percentages are to be read as $\pm$ 2% (95% probability). To obtain optimal counting efficiency, distances between adjacent points in the counting ocular were made larger than the largest dimensions of simple packing voids as present in thin sections of the basic fabrics (Van der Plas & Tobi, 1965). The size of a counted pore, below one of the points in the ocular, was determined by measuring the smallest possible intergranular distance in the pore through the point because the shortest dimensions in pore systems (their "necks") are most significant in determining resistance to water movement. Marshall (1958) introduced a model for calculating widths of "pore-necks" in sandy soils. Pore size distributions were indirectly derived from moisture retention data. Total porosity is divided into n small classes (see equation [1]), each class being represented by a representative pore size $r_n \cdot (r_1 > r_2 > r_3 \ldots r_n)$. The radius of the circular "neck" area r_t is then given by

$$r_t = \sqrt{e \cdot n^{-2}[r_1{}^2 + 3r_2{}^2 + 5r_3{}^2 + \ldots(2n-1)r_n{}^2]} \qquad [2]$$

where $e =$ porosity as measured with physical methods.

A similar procedure has to be followed when pore size distributions are obtained by direct microscopic measurement. A relatively large pore observed in a thin section may become very small in the soil material adjacent to the thin slice of soil represented in the section. This applies particularly to simple packing voids occurring in basic structures.

Pore size distributions, determined by morphometric analysis, were used to calculate K values for the basic fabrics using the pore "neck" concept similar to equation [5] of Marshall (1958), but given in the notation of Green and Corey (1971):

$$k(\theta)_i = (K_s/K_{sc}) \cdot \frac{(ep/n^2)}{8} \sum_{j=i}^{m} \left[(2j + 1 - 2i)r_j{}^{-2}\right] \quad (i = 1,2,\ldots m) \quad [3]$$

where $k(\theta)_i =$ the calculated k (intrinsic permeability) for a specific water content θ_v (cm³/cm³) in cm² (to be transformed into centimeters per day. Numerically $K = 1$ cm/day $\approx k_i = 10^{-10}$ cm²) $\cdot$ $e =$ total porosity determined with physical methods. The other symbols were defined previously.

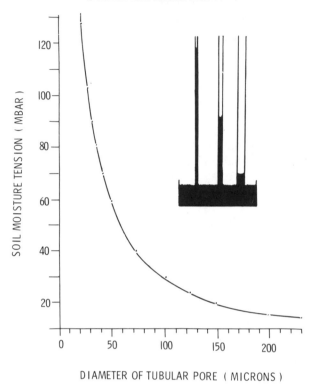

Figure 4. Relationship between soil moisture tension and tubular pore diameter according to $r = 2\delta/\rho gh$ (see text).

In basic fabrics containing plasma, porosities determined by point count are much lower than those physically measured (Bouma, 1969) because small packing pores between plasma grains are invisible in transmitted light in a thin section of 20 μm thickness. Therefore, total porosity was determined physically and the "morphological" porosity was divided into m classes, each with the same size as an n class, representing the largest pore classes in the sample. For example, with the physical porosity of a sample at 40% and $n = 20$, each class would represent 2% of the pore volume. If the morphological point count yielded a porosity of 20%, m was 10 and the morphologically determined pore size distribution from which K values had to be derived applied to θ_v from 40 to 20 cm³/cm³. It would also be possible to determine the relationship between K and a derived soil moisture pressure by calculating corresponding h values for every observed r value according to: $r = 2\delta/\rho gh$ (Marshall, 1958) (see Fig. 4) in which δ = surface tension of water, ρ = density of water, and g = gravitational constant. Then, total physical pore volume would not be needed in the calculations. Finally, r_j = average pore size in each small class; $j = 1, 2, \ldots .m$, as directly read from Fig. 5, which gives the pore size distribution as determined by microscopic count.

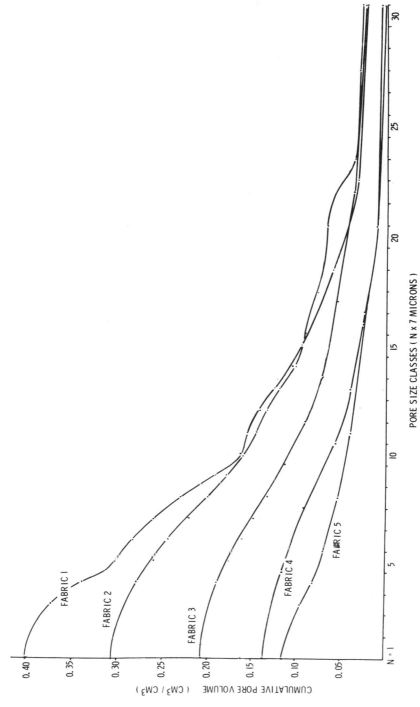

Figure 5. Pore size distribution of five basic structures determined by point-count procedures in thin section.

SECONDARY STRUCTURES

Pedality and occurrence of transpedal channels were described with methods discussed on p. 78–81. Here, a method will be explored to quantify patterns of interpedal planar voids between accommodated peds. Flow through plane slits can be described by physical equations that relate pore size to moisture flow at defined hydraulic gradients. For a plane slit of unit length and width d (Childs, 1969),

$$Q/t = [(\rho g d^3)/12\eta] \cdot \text{grad } \phi \qquad [4]$$

where Q/t = the amount of liquid, with viscosity $\eta (g \text{ cm}^{-1}\text{sec}^{-1})$, conducted per unit time and length ($\text{cm}^2\text{sec}^{-1}$); ρ = density of liquid ($g \text{ cm}^{-3}$); g = gravitational constant (cm sec^{-2}); and grad ϕ = hydraulic gradient (cm/cm). Assuming that only planar voids contribute to pore volume in a soil body which has n parallel slits per unit area each of width d, one finds: $K_{\text{sat}} = g\rho f d^2/12\eta$, in which f = total porosity occupied by these pores = $n \cdot d$ (Childs, 1969). In natural soil materials only very small fractions (1–2%) of total pore volume are occupied by planar voids. Fine voids in the basic fabric that contribute only negligibly to flow, account for a high percentage of total porosity (see Tables 6 and 7). Therefore, this equation for K_{sat} was not used and flow per unit area is described by

$$v = (n/S)(Q/t) = (ng\rho d^3/12\eta \cdot S) \cdot \text{grad } \phi \text{ and } K = ng\rho d^3/12\eta \cdot S \qquad [5]$$

where S = top surface area of measurement in a soil peel (cm^2) in which n cm of planar voids occur, and K (cm/sec) is derived at grad ϕ = 1 cm/cm. For tubular pores we find that flow rate and pore size are related as follows (Childs, 1969):

$$Q/t = (\rho g \pi r^4/8\eta) \cdot \text{grad } \phi \qquad [6]$$

where Q/t = the amount of liquid, with viscosity η, conducted per unit time (cm^3/sec), and r is tube radius (cm). With n tubes per cm^2, it follows

$$v = n(Q/t) = (n\rho g \pi r^4/8\pi) \cdot \text{grad } \phi$$

If these tubes are the only pores, $n\pi r^2 = f$, where f = total porosity. Then

$$v = (\rho g r^2 f/8\eta) \cdot \text{grad } \phi \text{ and } K = (\rho g r^2 f/8\pi) \text{ (Childs, 1969)}.$$

When scattered large channels occur in a porous soil mass, use of f may be confusing since channels contribute only a small fraction (f) to total porosity (see Table 6). If a soil horizon has n channels of radius r per S cm^2 horizontal soil surface, the following relationship applies:

A PLANAR-VOID STRUCTURE MODEL

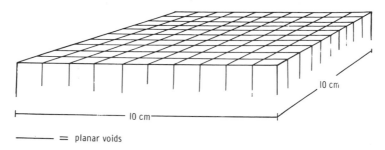

——— = planar voids

A TUBULAR-VOID STRUCTURE MODEL

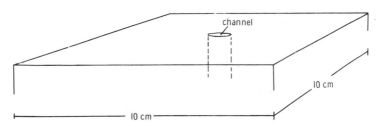

Figure 6. Horizontal sections through a tubular and a planar void model of soil structure.

$$v = (n/S) \, (Q/t) = (n\rho g \pi r^4/S \cdot 8\eta) \cdot \text{grad } \phi.$$

For grad $\phi = 1$ cm/cm we find K (cm/sec)

$$K = n\rho g \pi r^4/S \cdot 8\eta \qquad\qquad [7]$$

where v = the measured rate of infiltration into soil surface area S at saturation and at a given hydraulic gradient.

This approach can be illustrated by calculating K_{sat} values considering horizontal sections through abstract models of "soil" (Fig. 6). The upper figure shows a surface of 100 cm^2 occupied by square blocks and mutually separated by a distance d. The blocks are impermeable and water moves only along the planes. The lower figure shows the same surface with one tubular pore (channel). Water can only move through this pore. K values for these models were calculated for different sizes of pores (using equations [5] and [7], derived from [4] and [6] which are graphically represented in Fig. 7), assuming that planes and channels extend vertically downwards (see Table 6). "Effective planar widths" (d^1) were calculated for the four secondary fabrics, following this procedure in reverse by measuring the length of natural planar voids in soil peels and using a measured K_{sat} value.

It is assumed that movement of water occurs along planes only and that planar voids have one "effective width" and extend downwards into the soil for a distance of 12 cm in this case (Bouma and Hole, 1971).

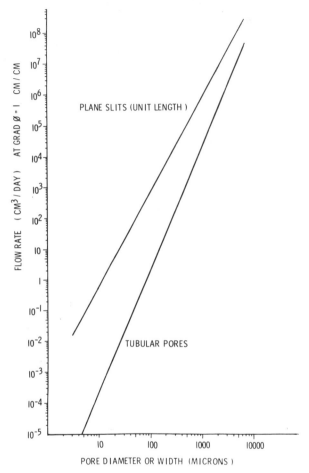

Figure 7. Relationships between pore size and flow rate, calculated for tubular and planar voids, according to equations [6] and [4].

$$d^1 = \sqrt[3]{(K \cdot 12\eta \cdot S)/(L \cdot \rho \cdot g)} \qquad [8]$$

where S = top surface area of the soil peel, d^1 = effective planar width, and L = total length of planar voids in soil peel. Other terms were explained earlier.

However, planar voids usually will not be vertically continuous in a soil sample that is several times larger than the average ped size. Therefore, the following procedure, using a method to estimate vertical plane continuity, was applied to calculate K_{sat} of the four secondary structures with a planar void model.

The procedure had the following steps:

1) A picture was made of soil peels from horizontal and vertical sections of the horizon to be characterized. Peels were large enough to represent at least 25 primary peds.

Table 5. Morphological and physical characteristics of six basic structures (Fig. 2).

					Physical data				
	Morphological data (air-dry sample)				Porosity		Bulk density		
No.	Porosity	Plasma	Skeleton grains	Related distribution pattern	Air-dry	Satu-rated	Air-dry	Satu-urated	COLE
	vol. %				%				
1	40.0	0.0	60.0	Granular	36.0	36.0	1.68	1.68	0
2	31.0	12.0	57.0	Intertextic	33.7	35.0	1.75	1.70	0.010
3	20.5	21.5	58.0	Intertextic	33.0	39.9	1.77	1.59	0.035
4	14.0	33.0	53.0	Intertextic→agglom-eroplasmic	26.6	37.6	1.95	1.66	0.058
5	11.3	48.7	50.0	Agglomeroplasmic	24.9	36.7	2.01	1.69	0.060
6	0.0	90.0	10.0	Porphyroskelic	29.8	39.9	1.93	1.65	0.055

2) Tracings were made of the pictures, showing patterns of natural planar voids (Fig. 3). Total lengths of planar voids were measured from pictures of horizontal peels with a map measurer, and ped sizes were determined by estimating (for every exposed ped) the diameter of a circle with a similar surface. Then average ped size was determined.

3) Distribution of widths of planar voids was determined in air-dry horizontal soil peels with a ribbon-count procedure (see Van der Plas, 1962; Brewer, 1964, p. 50). Line or point-count procedures were not suitable because the surface of planar voids was only a very small fraction of the total peel surface. The ribbon-count procedure implies measurement of numbers and sizes of individuals in the soil peel with a binocular microscope at a magnification of 30X in this case. The microscope was mounted so that the field of vision, including a measuring scale, could be moved over the peel surface. A small observational ribbon 60 μm wide was projected on the peel. Every

Table 6. Hydraulic conductivities and porosities of planar and tubular pore models of soil structure (Fig. 6).

Planar void model			
Size of blocks	Width of planar voids	Porosity	K
		%	cm/day
1 cm^2	10μ	0.2	1.5
	50μ	1.0	180
	100μ	2.0	1,440
4 cm^2	10μ	0.1	0.7
	50μ	0.5	90
	100μ	1.0	720
	1,000μ	10	719,712

Tubular pore model			
	Diameter of channel	Porosity	K
		%	cm/day
	100μ	0.8×10^{-4}	0.02
	200μ	0.3×10^{-3}	0.34
	500μ	0.2×10^{-2}	14.6
	1,000μ	0.8×10^{-2}	211.2
	4,000μ	0.13	54,130.5

planar void that crossed the ribbon was counted and the smallest observed dimension was recorded. In each peel at least 300 planar voids were counted and grouped in size classes corresponding to scale units.

4) A "representative plane length l_i" was calculated for each size class

$$l_i = L(p_i/p_t)$$

where L = total length of planar voids in an air-dry peel; p_i = number of planes in a certain size class i; i = 1,2,3. . .n; and p_t = total number of planes counted. Plane widths were measured in air-dry soil peels. COLE values (Grossman et al., 1968) were determined for primary peds to estimate volume changes of air-dry peds upon saturation and resulting decreases in plane width. If average ped size is 10 mm, a COLE value of 3% will increase this to 10.3 mm. In a soil material with accommodated peds, the width of all planar voids will then be 300 μm less. Planes smaller than 300 μm in the air-dry peel will close.

5) Calculated K values are to be compared with values measured *in situ* with the double-tube method. Simulation techniques have shown that measurement involves a soil sample with a height of about $2R_c$, which is 12 cm (Bouwer, 1962). In pedal soil materials with peds much smaller than 12 cm, K_{sat} will be determined by the smallest voids in the interpedal void system. A model is necessary to predict sizes of such "planar necks" in the flow system. The hypothesis used is that interconnected open planar voids govern the flow process. With an average vertical ped size of v cm, there are $12/v$ layers of peds. Swelling resulted in a reduction of the number of classes from n to $n - x$, where x is the largest pore size class closed by swelling (all classes between 1 and x are closed too). Chances of a size class of planes to be vertically continuous can be estimated as they are proportional to the l_i values for each open class considering a structure model with $12/v$ layers of peds. For example, open planes, interconnected through the layers of peds according to the hypothesis, are of size class $(n - x + 1)$ or larger. Chances to be of class $(n - x + 2)$ or larger (excluding class $n - x + 1$) are lower. Suppose l_i values for the classes are l_{n-x+1}, l_{n-x+2}, l_{n-x+3} . . ., etc., with total length of open planes after swelling l_t. Considering the contact between the first and second layers of peds, the probability of a plane of class $(n - x + 2)$ to be connected with a pore similar or larger in size than itself is proportional to the l_i values involved: $(l_t - l_{n-x+1})/l_t$. A similar expression can be derived for the first three layers of peds. Finally, probability that planes of size class $(n - x + 2)$ are connected throughout the 12-cm long sample by planes either of its own size or larger is

$$[(l_t - l_{n-x+1})/l_t]^{(12/v)-1}.$$

A similar calculation is made for the class $(n - x + 3)$ etc. Only if such probabilities are higher than Z% are they considered statistically significant, and corresponding pore classes are included in the calculation of conductivity

(Z was arbitrarily chosen as 5%). Assume that class $n - x + s$ represents the last class to have a probability higher than Z% of being continuous into a similarly sized or larger pore class. Now only a fraction p of open pore length (where $p = l_{n-x+1} + l_{n-x+2} + \ldots l_{n-x+s}$) contributes to conductivity. According to the model, flow into voids of size class $n - x + s + 1$ or larger will have to pass voids of size class $n - x + 1$ or larger while moving through the 12-cm high sample. The probability that flow occurs through size class $n - x + 1$ is highest. Therefore, l_i values of all pores larger than class $n - x + s$ are assigned to class $n - x + 1$. Two l_i values are given in Table 7 for this class, one directly calculated for the class and the second as a rest factor. The final equation for the calculation of K, using equation [5] for each of the size classes, then becomes

$$
K = 86400 \left\{ \left[\sum_{i=n-x+1}^{i=n-x+s} \left(\frac{\rho g d_i^3}{12\eta} \right) \times \frac{l_i}{S} \right] \right.
$$
$$
\left. + \frac{\rho g d^3_{n-x+1}}{12\eta} \times \frac{l_t - (l_{n-x+1} + l_{n-x+2} + \ldots l_{n-x+s})}{S} \right\} \qquad [9]
$$

where K = the hydraulic conductivity (cm/day), d_i = the average width (cm) of a planar void in a size class i, and S = top surface area of the soil peel. The rest of the terms are explained in the text.

RESULTS AND DISCUSSION

Basic Fabrics

The series of basic fabrics (Fig. 2) shows different types of related distribution patterns of plasma and skeleton grains ranging from granular to porphyroskelic (Brewer, 1964) (Table 5). Soil porosity, observed in thin section, decreases as the plasma content increases, whereas total porosity as determined by physical methods, stays more or less constant (Fig. 5, Table 5). Small pores between clay domains and fine silt-sized grains are not observed as individuals in a thin section with a normal thickness of around 20 μm. However, such pores constitute the larger part of soil porosity. Swelling of soil upon wetting has the effect of altering pore sizes. Air-dry samples were used for thin sectioning. COLE values were determined to estimate linear swelling of dry soil upon saturation (Grossman, 1968). Though swelling leads to a considerable increase in total clod volume, effects at the microlevel are small. Plasma bridges between skeleton grains have a thickness varying between approximately 20 and 70 μm that increase upon swelling to 21 and 74 μm, respectively. This increase is less than the scale unit of measurement (7 μm), and therefore, effects of swelling were neglected. The effect is stronger in fabric 5, which is not very suitable for this type of analysis because of its relatively high plasma content and low K_{sat}. Hydraulic con-

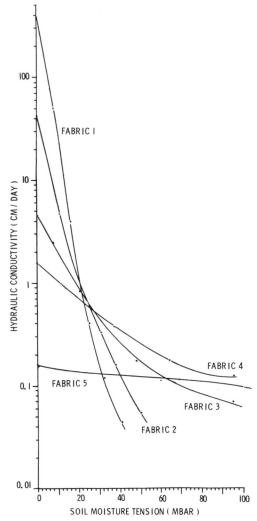

Figure 8. Hydraulic conductivity characteristics for five basic structures.

ductivity and moisture retention characteristics were measured in soil materials represented by samples 1 through 6 with increasing content of soil plasma.

The following physical effects were associated with increasing plasma content:

1) K_{sat} values decreased strongly because of formation of small pores at the points of contact and between skeleton grains (Fig. 8).

2) The decrease of K, with increasing soil moisture tension, was slower. At a moisture tension of around 25 mbar, K values of all fabrics, except 5, were similar although differences in K_{sat} were large (Fig. 8).

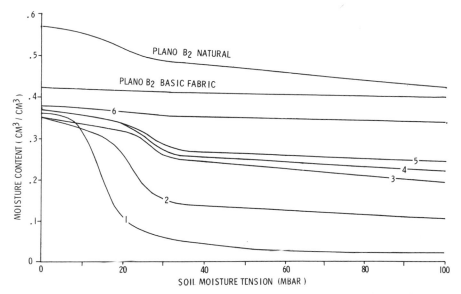

Figure 9. Moisture retention data for six basic structures and the B2 horizon of the Plano silt loam. The curve for the basic fabric of this horizon was obtained from puddled, air-dried, and remoistened soil material with simple packing voids only.

3) Moisture retention by the fabrics increased while desorption of water occurred less abruptly at increasing tensions (Fig. 9).

Moisture retention data and point-count results were used to calculate K values of the fabrics (Fig. 10). Reasonable agreement is found between measured and calculated values. But large matching factors (Green & Corey, 1971) were necessary in both methods to derive the calculated values (K_s/ K_{sc} in Fig. 10). At a higher clay content (sample 5), neither method predicted K well. A somewhat better agreement between calculated and measured values would have occurred had the curves been matched at a water content of 0.3 cm³/cm³, rather than at saturation (Green & Corey, 1971). However, the calculation method looses much of its merit when unsaturated K values have to be determined for curve-matching purposes. Therefore, use was made only of K_{sat} values which are relatively easy to measure. The physical method is based on the assumption that water extraction from a soil sample, after application of a certain pressure or suction, results from emptying pores within a certain size range that is described by a capillary model (Fig. 4). However, moisture extraction from clayey materials may be achieved by a decrease in total volume of the soil sample (see Table 5 with COLE values) without creating empty voids. In this case the principle of the method does not apply. The morphologic method does not work well when porosity (observed in thin section) is only a small fraction of the total porosity, as measured with physical methods. Then, observed relatively large pores dominate in the calculation procedure, whereas conductivity is governed

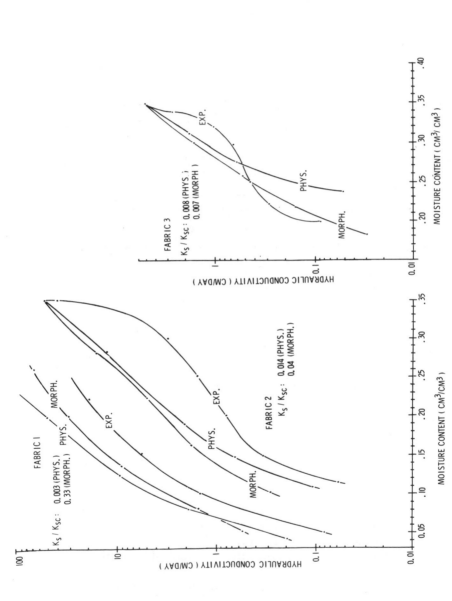

Figure 10. *K* values of five basic fabrics and the B2 of the Plano silt loam, experimentally determined (Exp.) and calculated with physical (Phys.) and morphological (Morph.) methods, including matching factors (K_s/K_{sc}) (continued on next page).

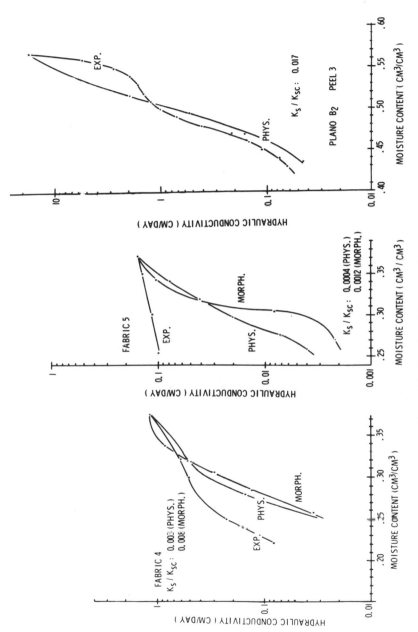

Figure 10. Continued from previous page.

Table 7. Calculation of K_{sat} for four secondary soil structures (Fig. 3).

Soil type	Horizon	Average ped size		COLE	L	l_t	Open pores after swelling		K calculated planar model		K measured	Effective planar width d'	Planar porosity	K calculated channel model
		Horizontal	Vertical				Size* class	l_i	Per class	Total				
		cm	cm		cm	cm	μm	cm	cm/day	cm/day	cm/day	μm	%	cm/day
Tama silt loam (No. 1, Fig. 3)	B3	1.50	2.0	0.030	127	48.6	36 96	15.6+20 9	4.0 19.9	23.9	22	39	0.2	1,500
Plano silt loam (No. 3)	B2	0.91	1.7	0.030	157	26.5	37 97	9.8+7.7 9.0	2 20.7	22.7	20	38	0.2	25,000
Plano silt loam (No. 4)	B3	1.90	2.5	0.028	86	18.4	18 78 138	3.4+8.1 5.1 1.8	0.1 6.8 11.8	18.7	12	36	0.1	21,650
Oshkosh clay (No. 5)	B2	0.76	1.5	0.029	107	23.0	32 92	5.9+10.9 6.2	1.4 6.2	7.6	6	29	0.1	1,390

* One microscopic scale unit corresponded here to 60 μm; larger magnifications are recommended for future work.

by smaller pores that are not observed as individual pores in the thin section (Bouma, 1969).

K_{sat} values of basic fabrics with a relatively high content of plasma are generally low (K_{sat} of fabric 5 was 1 mm/day and that of fabric 6 was unmeasurably low). The conductivity of soils with such basic fabrics is mainly determined by interpedal planar voids and transpedal channels that form the structural porosity. These pores will be discussed in the section on secondary structures.

Secondary Structures

Moisture retention data were used to calculate K values for the B2 of the Plano silt loam (Green & Corey, 1971). Good agreement was found with the experimental values (Fig. 10), although a considerable matching factor was necessary. Morphometric point-count results were not used to calculate K values because of the high cost involved as many thin sections would be necessary to represent the heterogeneous soil horizon. Occurrence of planar voids and channels in soil materials can result in high K_{sat} values even though they contribute minimal fractions to soil porosity (Table 6). Methods to estimate K_{sat} which are based on relative *volumes* of pore size classes, either determined micromorphologically with point counts or physically by moisture retention measurements, tend to be very insensitive because of the experimental error involved. This explains the necessity for use of large matching factors. An alternative is to count the number or length of specific pore types in a certain soil surface discussed in this paper.

Calculated values of d' (effective planar width, equation [8]), varied within a narrow range of 29 to 39 μm (Table 7). This could point to the feasibility of a method where total planar-void length L is measured in a certain soil surface area and K is calculated according to equation [8] after estimating an "effective planar width d'". The calculated values of K_{sat} for four secondary soil structures according to the plane-slit structure model, equation [9], were close to those measured with the double-tube method in the field (Table 7). This model applies to structures with accommodated, relatively large peds that should have a small range in horizontal size so as to make the corrections for swelling applicable. Flow in the model is supposed to be occurring only along planar voids, open after swelling, that constitute a system with interconnected planes. Since all basic structures of the four soil horizons had a porphyroskelic-related distribution pattern (like type 6 in Fig. 2), it can be assumed that water movement through these very fine basic structures is negligible. Relatively large, but isolated and vertically discontinuous voids such as vughs and chambers will not contribute to the conductivity of a soil layer about 12 cm thick. However, channels may be vertically continuous for quite a distance. Worm or ant channels, for example, may extend from the soil surface downwards to a depth of 15.24 dm (5 feet) (Baxter and Hole, 1967). The planar structure model was nevertheless used as a hypothetical model because

1) Illuviation and other cutans and skeletans on ped faces strongly indicate movement of liquid through planar voids. A picture of a ped surface and a ped interior of the B2 of a Plano silt loam (Fig. 11), taken with the Scanning Electron Microscope, shows that ped surfaces were partly sealed in contrast to the open ped interior, although no ped-illuviation cutans were observed in thin section. Similar examples were given by Lynn and Grossman (1970). Moreover, ped faces in a freshly exposed soil are often wet, while interiors of peds are only slightly moist.

2) Flow rates were calculated for the four studied soil horizons, using the counted number of channels on exposed horizontal sections, in the field (Table 3) and equation [7], applied to each size class followed by summation.

Calculated K values for the four soil horizons, based on the occurrence of channels only, are in the last column of Table 7. Values are unrealistically high particularly when channels larger than 2 mm are present. This is mainly caused by the lack of a suitable pore-continuity model (like the one used for interpedal planar voids) for channels occurring in natural soil. Very high K_{sat} values may be measured when large channels are vertically continuous throughout a relatively small soil core that is used to determine K_{sat} in the laboratory (Klute, 1965). However, in a natural profile chances are that such a large pore, if in contact with tension-free water, will fill up and drain through surrounding smaller, but more numerous planar or packing voids. Obviously, more research is needed to establish the hydraulic function of channels in the natural soil fabric.

3) The calculated values were compared with those determined by the double-tube method in the field. This measurement involves both horizontal and vertical flow components. Planar void patterns extend in all directions, whereas channels, notably the larger ones, tend to be more vertically oriented. Agreement between calculated and measured values is partly a result of the method used.

CONCLUSIONS

The purpose of using morphometric techniques should not be to reproduce results that can more easily be obtained by physical methods, but to use specific advantages inherent to such techniques, such as distinction and measurement of different types of pores. Planar voids and channels, though contributing only small fractions to total pore volume, strongly affect the conductivity of soil materials. Morphometry can be used to construct simple but relevent models of soil structure with planar-voids, channels, or packing voids from which K may be estimated using basic physical equations relating pore size to conductivity. In a pore system with varying sizes, conductivity is governed by the smallest pores in the system. Therefore, hypothetical models are necessary to predict pore continuity in soil materials from pore measurements that are necessarily made in two-dimensional sections. A continuity model for interpedal planar voids was introduced.

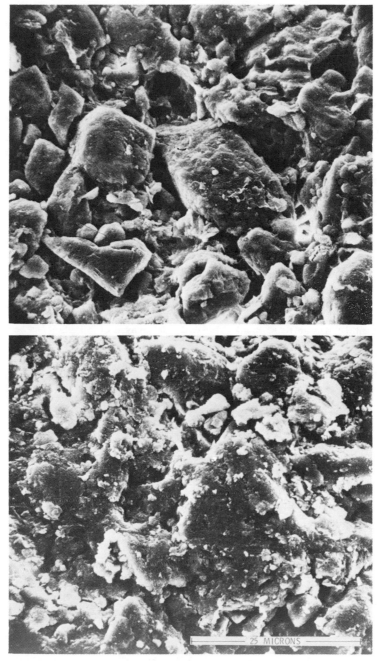

Figure 11. Ped interior (upper) and ped surface in a B2 horizon of a Plano silt loam. Picture was made with a Scanning Electron Microscope (Photo courtesy of Dr. E. B. Sachs, USDA Forest Products Laboratory, Madison, Wis.).

The concept of soil structure should apply not only to aggregated, but also to nonaggregated soil materials because they have characteristic basic structures that strongly affect their hydraulic behavior.

Pore size distributions, obtained directly by point-count procedures in thin sections of soil materials, can be used to predict K values, as has been demonstrated for a series of basic structures. Results were compared to those obtained by a physical method using moisture retention data to indirectly characterize the pore size distribution. Both procedures yielded comparable results but did not predict K directly and needed large "matching factors" derived from actual measurement of K_{sat}. The physical method is preferable as it is simpler and cheaper.

Morphological study of soil materials with the purpose of relating soil structure to hydraulic conductivity involves observation of basic, matric, secondary, and tertiary soil structures in sufficient detail to obtain an informative, yet readable and reproducible description. Relationships between this type of data and hydraulic conductivity characteristics, as explored in this paper, can be used to improve field estimates of K.

LITERATURE CITED

Anderson, D. M., and R. R. Binnie. 1961. Modal analysis of soils. Soil Sci. Soc. Amer. Proc. 25:499–503.

Baumgart, H. D. 1967. Die Bestimmung der Wasserleitfähigkeit kf von Böden mit tief liegender Grundwasseroberfläche. Mitteilungen aus dem Institut für Wasserwirtschaft und Landwirtschaftlichen Wasserbau der Technischen Hochschule Hanover. 275 p.

Baxter, P. F., and F. D. Hole. 1967. Ant (*Formica cinerea*) pedoturbation in a prairie soil. Soil Sci. Soc. Amer. Proc. 31:425–428.

Blake, G. R. 1965. Particle density. *In* C. A. Black et al. (ed.) Methods of soil analysis, Part 1. Agronomy 9:371–373.

Boersma, L. 1967. Field measurement of hydraulic conductivity above a water table. *In* C. A. Black et al. (ed.) Methods of soil analysis, Part 1. Agronomy 9:235–252.

Bouma, J. 1969. Microstructure and stability of two sandy loam soils with different soil management. Agr. Res. Rep. 724, p. 110. Pudoc Wageningen, The Netherlands.

Bouma, J., D. I. Hillel, F. D. Hole, and C. R. Amerman. 1971. Field measurement of unsaturated hydraulic conductivity by infiltration through artificial crusts. Soil Sci. Soc. Amer. Proc. 35:362–364.

Bouma, J., and F. D. Hole. 1965. Soil peels and a method for estimating biopore size distribution in soil. Soil Sci. Soc. Amer. Proc. 29:483–485.

Bouma, J., and F. D. Hole. 1971. Soil morphology and hydraulic conductivity of adjacent virgin and cultivated pedons at two sites: A Typic Argiudoll (silt loam) and a Typic Eutrochrept (clay). Soil Sci. Soc. Amer. Proc. 35:316–319.

Bouwer, H. 1961. A double tube method for measuring hydraulic conductivity of soil *in situ* above a water table. Soil Sci. Soc. Amer. Proc. 25:334–339.

Bouwer, H. 1962. Field determination of hydraulic conductivity above a water table with the double tube method. Soil Sci. Soc. Amer. Proc. 26:330–335.

Bouwer, H., and H. C. Rice. 1964. Simplified procedure for calculation of hydraulic conductivity with the double tube method. Soil Sci. Soc. Amer. Proc. 28:133–134.

Bouwer, H., and H. C. Rice. 1967. Modified tube diameters for the double tube apparatus. Soil Sci. Soc. Amer. Proc. 31:437–439.

Brasher, B. R., D. P. Franzmeier, V. Valassis, and S. E. Davidson. 1966. Use of Saran resin to coat natural soil clods for bulk density and water retention measurements. Soil Sci. 101:108.

Brewer, R. 1964. Fabric and mineral analysis of soils. John Wiley & Sons, New York. 470 p.

Buol, S. W., and D. M. Fadness. 1961. A new method of impregnating fragile material for thin sectioning. Soil Sci. Soc. Amer. Proc. 25:253.

Chayes, F. 1956. Petrographic modal analysis. John Wiley & Sons, New York.

Childs, E. C. 1969. The physical basis of soil water phenomena. John Wiley & Sons, New York.

Gardner, W. R. 1970. Field measurement of soil water diffusivity. Soil Sci. Soc. Amer. Proc. 34:832-833.

Green, R. E., and J. C. Corey. 1971. Calculation of hydraulic conductivity: A further evaluation of some predictive methods. Soil Sci. Soc. Amer. Proc. 35:3-8.

Grossman, R. B., B. R. Brasher, D. P. Franzmeier, and J. L. Walker. 1968. Linear extensibility as calculated from natural-clod bulk density measurements. Soil Sci. Soc. Amer. Proc. 32:570-573.

Hillel, D. I., and W. R. Gardner. 1970. Measurement of unsaturated conductivity and diffusivity by infiltration through an impeding layer. Soil Sci. 109:149-153.

Jager, A., and W. J. M. van der Voort. 1966. Collection and preservation of soil monoliths. Soil Survey Papers No. 2, Soil Survey Institute, Wageningen, The Netherlands.

Jongerius, A. 1957. Morfologische onderzoekingen over de bodemstructuur. Verslagen Landbouwkundige Onderzockingen no. 63-12. Wageningen, The Netherlands. 93 p.

Klute, A. 1965. Laboratory measurement of hydraulic conductivity of saturated soil. *In* C. A. Black et al. (ed.) Methods of soil analysis, Part 1. Agronomy 9:210-221.

Kubiena, W. L. 1938. Micropedology. Collegiate Press Inc., Ames, Iowa.

Kunze, R. J., G. Uehara, and K. Graham. 1968. Factors important in the calculation of hydraulic conductivity. Soil Sci. Soc. Amer. Proc. 32:760-765.

Lynn, W. C., and R. B. Grossman. 1970. Observations of certain soil fabrics with the scanning electron microscope. Soil Sci. Soc. Amer. Proc. 34:645-648.

Marshall, T. J. 1958. A relation between permeability and size distribution of pores. J. Soil Sci. 9(1):1-8.

Slager, S. 1966. Morphological studies of some cultivated soils. Agr. Res. Rep. 670. Pudoc Wageningen, The Netherlands.

Soil Survey Staff. 1951. Soil survey manual. USDA Handbook 18. 503 p.

Van der Plas, L. 1962. Preliminary note on the granulometric analysis of sedimentary rocks. Sedimentology 1:145-157.

Van der Plas, L., and S. Slager. 1964. A method to study the distribution of biopores in soils. p. 411-419. *In* A. Jongerius (ed.) Soil micromorphology. Elsevier, Amsterdam.

Van der Plas, L., and A. C. Tobi. 1965. A chart for judging the reliability of point counting results. Amer. J. Sci. 263:87-90.

Water Retention and Flow in Layered Soil Profiles[1]

6

D. E. MILLER[2]

ABSTRACT

In general, any profile discontinuity that affects pore size distribution will decrease water movement across the discontinuity boundary compared with a uniform profile. If the profile contains a layer less permeable to water than the soil above, water can be transmitted through the soil more rapidly than through the layer and may accumulate above the layer. A coarse layer has a high saturated conductivity, but it ceases to transmit significant amounts of water at relatively low suctions. Thus, the suction in soil above a coarse layer will be lower and the water content higher than in a nonlayered soil. The water retained in soil above a coarse layer is determined by the coarseness of the layer, depth to the layer, and desorption characteristics of the soil. The effects of coarse and slowly permeable layers are similar in that suction distributions are dominated by the nature and position of the layer, and water contents above the layer are related to soil desorption characteristics.

INTRODUCTION

The amount of water retained by a soil is affected by the physical characteristics of the wetted part of the profile. The water content at any place in a profile is due, in part, to the soil properties at that place, but it may be influenced by properties elsewhere in the profile (Richards, 1955). For example, the water content at a point above a layer in a profile is influenced by both the layer and the soil above, even though the layer may be some distance below the point in question.

In comparison with a uniform profile, any profile discontinuity that affects pore size distribution, such as a textural change, will result in decreased water movement. Water content above the boundary will be increased (Robins, 1959), if the soil has received sufficient water to allow drainage to occur beyond the depth in question. If the discontinuity is a layer of finer texture than the soil above, water will drain through the upper soil faster than through the fine-pored layer when the upper soil becomes wet enough. Water will then accumulate above the layer and positive hydraulic pressures may occur. If the discontinuity is a layer that is coarser than the soil above, the layer will not conduct significant amounts of water until

---

[1] Contribution from the Western Region, Agricultural Research Service, U. S. Department of Agriculture, in cooperation with the College of Agriculture, Washington State University, Pullman, Wash. Information paper.

[2] Soil Scientist, U. S. Department of Agriculture, Prosser, Washington.

many of the pores are filled with water. This will occur only at suctions much lower than those at which the pores in the soil above are filled. As a result, the soil water content will be greater when a coarse layer is present compared to a uniform soil. As the profile drains, the layer stops transmitting water at relatively low suctions, and the water content in the soil above the layer remains higher than in a nonlayered soil.

For this discussion, soils may be grouped into the four profile types discussed by Robins (1959) with the realization that there are many variations and gradations among them. These groups are: (i) deep soils, uniform in texture and structure; (ii) profiles with slowly permeable or nonpermeable layers, including soils underlain by bedrock; (iii) profiles in which soil is underlain by coarse materials; and (iv) profiles in which textural gradations occur over an appreciable depth range.

The water retention properties of deep, uniform soils are dominated by texture and to some extent by structure, providing the water table is deep enough that it does not influence water flow from the soil depth of interest. Since the subject of this paper is layered soils, these uniform profiles will not be discussed further.

With respect to the fourth group, Robins (1959) states that their behavior depends upon the depth and the intensity of the textural difference and the rate of change of texture within the profile. The soil water content above the transition zone will behave much as if the transition occurred abruptly at some depth within the zone. Accordingly, only groups two and three will be treated with major emphasis on group three.

Because soil water storage and availability is affected by many conditions other than soil profile, such as depth of wetting, plant vigor, rooting habit, and climatic factors, it will be assumed that the soil has been adequately wetted; and only the influence of the soil will be considered, except for minor references to the effect of evapotranspiration rate. The principles of water flow and retention in layered soils will be presented in general terms rather than being rigorously quantitative.

The effect of man-introduced profile layers, such as asphalt barriers, will also be discussed briefly.

PROFILES WITH LESS PERMEABLE SUBSOILS

There are many variations of the type of profile with less permeable subsoils. Slowly permeable subsoil layers may have formed in place, such as a high-clay B2 horizon. The less permeable layer may have resulted from an earlier deposition of soil material, such as lacustrine deposits that have been covered later by windblown soil. Some soils have more or less continuous, nearly horizontal, caliche layers. An extreme case is soil underlain by bedrock where the rock may or may not contain enough fractures to allow water to drain. Water that would normally drain in a deep soil is confined above the layer. A water table will result if the layer permeability is low enough and sufficient water is added to the surface. Unless an appreciable slope exists, the water will remain in the soil until it is lost by evapotranspiration and by slow drainage through the substrata. Inasmuch as the water would drain out of the root zone if the layer were not present, the layer greatly increases water storage capacity of the soil above the layer. Care must be exercised in irrigating these soils to prevent adverse effects of excess water and poor aeration. If the underlying layer is fractured rock, with the fractures capable of conducting some water, the condition becomes very similar to one of soil underlain by coarse materials.

A slowly permeable layer will transmit water slowly but for a long period of time after irrigation. Thus, the effective available water will be much higher with high evapotranspiration rates than with lower ones because the plant can extract water while the soil is draining. This water would be lost to deep drainage if plants were not present.

While the soil is wet above a layer, soil water suction will increase about 1 mbar/cm increase in elevation. The water content distribution will be determined by the ability of the soil to retain water at the imposed suction. In this regard, the effects of coarse and fine layers are similar in that the suction distributions are dominated by the nature and position of the layers, and water contents are then related to the soil desorption characteristics.

PROFILES WITH LAYERS OF COARSE MATERIALS

Alway and McDole (1917) observed that soil over a coarse layer retained more water after irrigation than a similar nonlayered soil. Gardner revived interest in this subject and produced a widely used film on water movement (Walter H. Gardner. 1960. Water movement in soil. A time-lapse motion picture of water movement in soils. Washington State Univ., Pullman.). Work has also been reported by Robins (1959), Eagleman and Jamison (1961), and Miller (1969).

The effect of a coarse layer on water retention is similar to that for a slowly permeable layer except that a coarse layer will not allow formation of positive water pressures above it. The layer will conduct water rapidly at suctions near zero and water from the overlying soil can easily escape, assuming that such flow is not restricted by such factors as a high water table or impermeable layers. At relatively low suctions, the coarse layer will not conduct water rapidly, and further drainage is restricted. Suction above the layer will increase nearly linearly with elevation, and hydraulic gradient within the soil will approach zero. The water contents are then dependent upon the desorption characteristics of the soil, as discussed in the previous section.

In brief, the suction distributions are determined by the characteristics of and depth to the coarse layer; and the water content distributions are then a function of the soil desorption characteristics. The major items to consider in evaluating the influence of coarse layers on water retention are (i) the coarseness of the layer (including the amount of fine materials), (ii) depth to the layer, and (iii) desorption characteristics of the soil.

Coarseness of the Layer

The effect of the coarseness of the layer on water retention can be illustrated by considering idealized profiles in which soil is underlain by sand of various size ranges. Unsaturated conductivity curves for some sand fractions are given in Fig. 1 (Miller, 1969).

If the upper limit of retained water is taken as that held in the soil when the drainage rate has decreased to about 0.01 cm/day, soil water suctions in the soil above a layer may be estimated from these curves if the hydraulic gradient within the sand layer is assumed to be near unity. Figure 1 illustrates that it makes little difference whether one selects 0.1 or 0.01 cm/day. The unsaturated conductivity curves are steep and the suctions are not greatly different for the two rates. Water cannot drain from the soil faster than the rate at which the coarse layer can transmit it. With the assumption of a hydraulic gradient of unity within the sand, the suction at the soil/sand interface when drainage from the soil essentially ceases will be as follows:

Diameter Range of Sand	Suction at Soil/Sand Interface
0.1 to 0.25 mm	70 to 80 mbars
0.25 to 0.5 mm	40 mbars
0.5 to 1.0 mm	20 mbars
1.0 to 2.0 mm	10 mbars

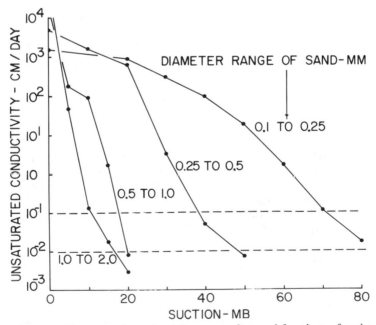

Figure 1. Unsaturated conductivity curves for sand fractions of various size ranges.

Table 1. Approximate suction distributions with depth in a soil underlain at 60-cm depth by sand of various size ranges.

| Soil depth | Suction distribution above the sand layer | | | |
| | Diameter range of sand, mm | | | |
	0.1 to 0.25	0.25 to 0.5	0.5 to 1.0	1.0 to 2.0
cm	mbars			
0	130	100	80	70
30	100	70	50	40
60	70	40	20	10

Table 2. Approximate water distributions with depth in a sandy loam soil underlain at 60-cm depth by sand of various size ranges. Values are estimated from suction data in Table 1 and desorption curve in Figure 2.

| Soil depth | Soil water distribution above the sand layer | | | |
| | Diameter range of sand, mm | | | |
	0.1 to 0.25	0.25 to 0.5	0.5 to 1.0	1.0 to 2.0
cm	volume %			
0	31	34	36	37
30	34	37	37	37
60	37	37	38	38

Suction within the soil above the layer will increase about 1 mbar/cm increase in elevation (the hydraulic gradient within the soil will be nearly zero). For example, if soil is underlain by these materials at a depth of 60 cm, the approximate suction distributions after drainage nearly ceases will be as given in Table 1.

If the desorption characteristics of the soil are known, then water contents can be estimated. For example, consider the desorption curve for a sandy loam soil as shown in Fig. 2 (Miller, 1969). This soil loses about 14% water by volume between 75 and 200 mbars, but very little at suctions below 75 mbars. Thus, the water content in 60 cm of soil over the finer sand

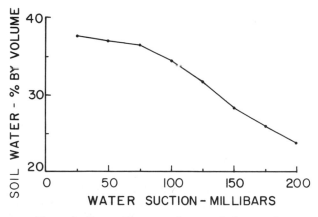

Figure 2. Desorption curve for a sandy loam soil.

Table 3. Total and available water in a sandy loam soil underlain at 60-cm depth by sand of various size ranges compared to the upper 60-cm of a deep uniform soil. Unavailable water is about 9% by volume.

	Diameter range of sand, mm				Uniform soil
	0.1 to 0.25	0.25 to 0.5	0.5 to 1.0	1.0 to 2.0	
Total water in upper 60 cm, cm	20.5	21.9	22.3	22.5	13.8
Available water in upper 60 cm, cm	15.1	16.5	16.9	17.1	8.4
Available water ratio, $\frac{\text{layered soil}}{\text{uniform soil}}$	1.80	1.96	2.01	2.04	

will vary with depth, but will be quite uniform in soil over the coarser sand as shown in Table 2.

The total water above the sand layers can be estimated from the data in Table 2. This soil, if uniform and deep, will drain to about 200 mbars within a few days after irrigation (Miller, 1969). The water content at 200 mbars suction will be about 23% by volume. The water retained under 15 bars suction will be about 9% by volume. Available water can be estimated for the layered soils and compared with nonlayered soils as has been done in Table 3. Thus, for this sandy loam soil, available water is approximately doubled if the underlying layer consists of 0.25 mm or larger material. As will be discussed later, the effect of the sand size depends upon the slope of the soil water desorption curve. A coarser soil with a steeper curve will have larger differences in water content among the sizes of underlying material, while a finer soil with a gentle slope to the curve may retain nearly the same amount of water for all of the layers.

If sufficient water has been applied to a uniform soil to wet below the root zone, the proportion that is lost as drainage will increase as the evapotranspiration rate decreases (Miller & Aarstad, 1971). This is because the uniform soil will drain for many days after irrigation if the soil water is not reduced by evapotranspiration. However, when a soil is underlain by a coarse layer, drainage essentially stops within a few days after irrigation. Thus, the increase in available water due to layering becomes relatively more important at low than at high evapotranspiration rates.

Depth to the Layer

Consider an idealized situation with the same sandy loam soil underlain at various depths by a sand that is 0.5 to 1.0 mm in diameter, so that the suction at the soil/sand interface will be about 20 mbars. As an approximation, suction will increase linearly with distance above the interface as discussed previously, so that the suction distribution will be as given in Table 4.

If the deep uniform soil drains to about 200 mbars, one would expect little effect from the deepest layer on water content at the surface (180 cm

Table 4. Approximate suction distributions with depth in a soil underlain at various depths by 0.5- to 1.0-mm sand.

	Suction distribution above the sand layer		
	Depth to layer, cm		
Soil depth	60	120	180
cm	mbars		
0	80	140	200
60	20	80	140
120		20	80
180			20

Table 5. Available water in a sandy loam soil underlain at various depths by 0.5- to 1.0-mm sand, compared to available water retained in similar depths of uniform soil.

	Available water in indicated depth range			
	Depth to layer, cm			
Soil depth	60	120	180	Uniform
cm	cm			
0 to 60	16.9	14.5	10.4	8.4
60 to 120	--	16.9	14.5	8.4
120 to 180	--	--	16.9	8.4

above the layer). Again referring to the desorption curve for this soil and considering available water, distributions can be estimated as shown in Table 5.

When the layer is near the surface, water in the soil above the layer is maintained in a position available to shallow-rooted crops. When the layer is deep (180 cm in the example), the water content at the soil surface is not affected appreciably but the amount of water available to a deep-rooted crop is much higher than in a nonlayered soil.

These estimations have ignored hysteresis effects. It is recognized that the desorption characteristics of the deeper soils will differ from those of the shallow soils, especially in the midpart of the profile because of the degree of wetting attained by the soil before it drained. At the surface and near the layer, desorption characteristics will be similar. Numerical solutions of the unsaturated flow equation, based on the technique of Hanks and Bowers (1962), give results that are in general agreement with those based on idealized profiles as discussed above (Miller, 1969).

Texture of Soil Overlying the Coarse Layer

Consider an idealized situation in which soils of various textures are underlain at a depth of 60 cm by a 0.5- to 1.0-mm sand. Desorption curves for three soils—a loamy sand, a loam, and a silt loam—are given in Fig. 3 (based on disturbed samples passed through a 1-mm sieve and saturated overnight before desorbing). The suction distribution above the layer will be similar

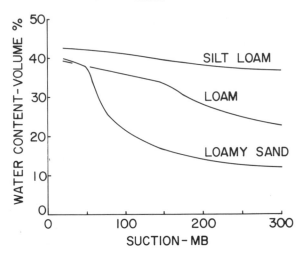

Figure 3. Desorption curves for silt loam, loam, and loamy sand soils.

in all soils and about the same as given in Table 1 for soil above this sand. In a deep, uniform profile the loamy sand will drain to about 125 mbars, the loam to about 200 mbars, and the silt loam to 300 mbars or more (D. E. Miller, unpublished data). Water contents corresponding to these suctions may be used as an estimate of the upper limit of available water in the soil and the 15-bar percentage as an estimate of unavailable water. Available water estimates for layered and uniform soils are given in Table 6. These estimates show that the largest increase in available water, compared to uniform soil, will occur when the overlying soil loses appreciable water as suction increases at some suction greater than what it will drain to over the sand layer. If the loamy sand were underlain by a finer material, such as the 0.1- to 0.25-mm sand, the increase in available water would be much less as many of the pores that remained full over the 0.5- to 1.0-mm sand would drain at the higher suctions that occur over the finer sand.

Field Observations

Field profiles vary from those that contain sharply defined layers with a pronounced influence on water retention to conditions where the layering has negligible effect. Attempts have been made to estimate the effects of natural layers on the water retention properties of the overlying soil.

Table 7 gives the unsaturated conductivities of samples from the coarse layers of several profiles together with the amount of fine material (< 0.25 mm diameter) in the samples (Miller, 1969). From these data, the following generalizations may be made:

    1) If the coarse material contains less than about 1% of < 0.1 mm material, soil over such material should not drain to suctions greater

Table 6. Available water in soils of various textures underlain at 60-cm depth by 0.5- to 1.0-mm sand.

	Available water in soil above layer		
	Loamy sand	Loam	Silt loam
	cm		
Above sand, cm	16.4	17.4	20.0
In 60 cm of uniform soil*, cm	6.7	11.7	16.7
Ratio: Layered			
Uniform	2.46	1.48	1.18

* Upper limit of available water is taken as that corresponding to a suction of 125 mbars in the loamy sand, 200 mbars in the loam, and 300 mbars in the silt loam.

than about 30 to 50 mbars at the interface. The expected suctions increase with the amount of $<$ 0.1 mm material, and at about 7% will be about 70 to 90 mbars.

2) If more than about 15% of the $<$ 0.1 mm material is present, it is probable that the suction distribution will be little affected by the layer and, thus, the water retention will be little affected.

MAN-INTRODUCED LAYERS

The above discussion has considered types of profile layering that may be encountered in the field. Also of interest are the attempts that have been made to obtain the benefits of profile layering by the introduction of a water flow barrier into the profile. Probably the most noteworthy are those studies in which flow barriers of asphalt have been placed at various depths in the profiles.

Erickson and co-workers (1968a; 1968b; Hansen & Erickson, 1969) noted that sandy soils with clay lenses in the subsoil had greater water holding capacity than when the lenses were absent. They attempted to achieve a similar increase in water-holding capacity by placing a layer of bentonite clay or plastic film in a sandy soil profile. Because of the difficulty in joining the material into a continuous barrier over a large area, the clay and plastic were abandoned in favor of liquid asphalt. They devised techniques whereby a

Table 7. Qualitative relation between unsaturated conductivities and the amounts of $<$ 0.1 and 0.1- to 0.25-mm fractions in coarse materials.

Source of layer material	Unsaturated conductivity at indicated suction (mbars)				Fine material in coarse layer	
	10	30	50	70	<0.1 mm	0.1 to 0.25 mm
	cm/day				%	
Model profiels	97.7	4.0	0.04	0.01	1.0	5.0
Ephrata loam	0.6	0.01	--	--	1.0	0.9
Timmerman sandy loam	4.6	0.14	0.06	0.03	3.9	2.6
Rupert sandy loam	24.1	5.43	0.84	0.14	6.8	13.2
Scooteney silt loam	7.0	2.64	1.57	1.29*	18.0	6.6

* Suction was 60 mbars.

continuous barrier of asphalt could be placed at the desired depth. The value of the barrier was demonstrated with a number of different crops and in a number of different areas with and without irrigation. Saxena, Hammond, and Lundy (1971) made similar investigations. In most of these studies, the asphalt barrier has about doubled the water holding capacity of the root zone of sandy soils with the greatest increase being just above the barrier. In Taiwan (Erickson, Hansen, & Smucker, 1968a), a fine sand planted to sugar cane drained to a suction of about 66 mbars within a few hours after irrigation. When an asphalt barrier was installed at a depth of 75 cm, the suctions above the barrier remained much lower. One month after irrigation, with the cane crop transpiring water, the suction just above the barrier was only 41 mbars. In Florida (Saxena et al., 1971), suctions over a barrier at a depth of 60 to 70 cm remained at less than 0.1 bar most of the season, compared with suctions often over 0.3 bar without the barrier. Water contents varied accordingly.

The barriers have increased productivity of soil in humid climates because water obtained by rainfall was retained in the root zone much longer than in the soil without barriers (Erickson et al., 1968a, 1968b; Saxena et al., 1971). In irrigated areas, the barriers have made possible significant savings in water by reducing the deep drainage losses following irrigation (Hansen & Erickson, 1969).

Unger (1971a, 1971b) was concerned with reducing evaporation losses of soil water; he attempted to achieve this reduction with surface and sub-surface layers of gravel. The gravel layers were 2.5 cm thick and were placed on the soil surface or at a depth of 5, 15, or 25 cm. Under conditions of natural evaporation at Bushland, Texas, there were trends toward greater water storage when the gravel layer was on the surface or at a depth of 5 cm compared with the control. Water storage was decreased when the layers were at 15 or 25 cm. These results were attributed to water flowing into the soil below the surface or shallow layer of gravel and thus being less subject to evaporation. With deeper gravel layers, most of the precipitation was retained above the gravel layer and was then positionally available for evaporation. In a somewhat similar study, Willis, Haas, and Robins (1963) found no effect on water storage from installation of a plastic film covering 90% of the soil at a depth of about 20 cm. Barriers resulting from plowing down a layer of plant material act similarly (Jamison, 1960).

These results substantiate the conclusions discussed earlier relative to the effect of profile layers on water retention. Water content at a flow barrier is increased and the effect decreases with height above the barrier. If the barrier is near the surface, the retained water is readily available for evaporation. If the barrier is deep enough, there will be little influence on the water content of the soil surface, but the stored water above the barrier will be increased compared to the similar soil without a barrier.

Miller and Aarstad (1972) attempted to achieve the benefits of an underlying coarse layer by plowing to a depth of about 100 cm with a moldboard plow. The soil, of the Hezel series, consisted of about 35 to 45 cm of loamy sand underlain by water-lain silty materials. They hoped to reduce suctions resulting from drainage by increasing the coarseness of the subsoil at plowing depth. The plow did not invert the profile but turned it about 135° so that the plowed soil consisted of alternating strips of silty and sandy soil at that angle. The plowing did increase the water holding capacity of the surface soil, but the increase was attributed largely to the change in soil texture of the surface with a relatively minor effect on the suction distribution after irrigation.

LITERATURE CITED

Alway, F. J., and G. R. McDole. 1917. Relation of water-retaining capacity of a soil to its hygroscopic coefficient. J. Agr. Res. 9:27–71.

Eagleman, J. R., and V. C. Jamison. 1961. The influence of soil textural stratification and compaction on moisture flow. Univ. Missouri Agr. Exp. Sta. Res. Bull. 784.

Erickson, A. E., C. M. Hansen, and A. J. M. Smucker. 1968a. The influence of subsurface asphalt barriers on the water properties and the productivity of sand soils. Int. Congr. Soil Sci., Trans. 9th (Adelaide, Aust.) I:331–337.

Erickson, A. E., C. M. Hansen, A. J. M. Smucker, K. Y. Li, L. C. Hsi, T. S. Wang, and R. L. Cook. 1968b. Subsurface asphalt barriers for the improvement of sugar cane production and the conservation of water on sand soil. Int. Soc. Sugar Cane Technol., Proc. 13th Congr. (Taiwan) Elsevier Publ. Co., Amsterdam. p. 787–792.

Hanks, R. J., and S. A. Bowers. 1962. Numerical solution of the moisture flow equation for infiltration into layered soils. Soil Sci. Soc. Amer. Proc. 26:530–534.

Hansen, C. M., and A. E. Erickson. 1969. Use of asphalt to increase water-holding capacity of droughty sand soils. Ind. Eng. Chem. Prod. Res. Develop. 8:256–259.

Jamison, V. C. 1960. Water movement restriction by plant residues in a silt loam soil Int. Congr. Soil Sci. Trans. 7th (Madison, Wis) I:464–472.

Miller, D. E. 1969. Flow and retention of water in layered soils. USDA Conservation Research Report No. 13: 28 p.

Miller, D. E., and J. S. Aarstad. 1971. Available water as related to evapotranspiration rates and deep drainage. Soil Sci. Soc. Amer. Proc. 35:131–134.

Miller, D. E., and J. S. Aarstad. 1972. The effect of deep plowing on the physical characteristics of Hezel soil. Washington State Univ. Circ. 556.

Richards, L. A. 1955. Water content changes following the wetting of bare soil in the field. Soil Sci. Soc. Florida, Proc. 15:142–148.

Robins, J. S. 1959. Moisture movement and profile characteristics in relation to field capacity. Int. Comm. Irrig. Drain. 8:509–521

Saxena, G. K., L. C. Hammond, and H. W. Lundy. 1971. Effect of an asphalt barrier on soil water and on yields and water use by tomato and cabbage. J. Amer. Soc. Hort. Sci. 96:218–222.

Unger, Paul W. 1971a. Soil profile gravel layers: I. Effect on water storage, distribution, and evaporation. Soil Sci. Soc. Amer. Proc. 35:631–634.

Unger, Paul W. 1971b. Soil profile gravel layers: II. Effect on growth and water use by a hybrid forage sorghum. Soil Sci. Soc. Amer. Proc. 35:980–983.

Willis, W. O., H. J. Haas, and J. S. Robins. 1963. Moisture conservation by surface or subsurface barriers and soil configuration under semiarid conditions. Soil Sci. Soc. Amer. Proc. 27:577–580.

Field Moisture Regimes and Morphology of Some Arid-Land Soils in New Mexico[1]

7

CARLTON H. HERBEL AND LELAND H. GILE[2]

ABSTRACT

This paper relates soil moisture in an arid region of southern New Mexico to precipitation, soil morphology, landscape position, and runoff. Matric potential of soil water on arid rangeland was determined from 1960–1970 with gypsum electrical resistance blocks using an ohmmeter. Soil texture, landscape position, and microrelief had a significant effect on soil moisture. Soil water potential at the 25-cm depth was between 0 and −15 bars an average of 40 days in a fine Haplargid that did not receive run-in, and an average of 212 days in a coarse-loamy Paleargid. With run-in, the fine Haplargid had an average of 82 days when the soil water was between 0 and −15 bars. In another fine Haplargid, there was an average of 166 days when the soil water was between 0 and −15 bars at the 25-cm depth, and 204 days at the 60-cm depth. At this site the depth of water infiltration was considerably increased by small depressions and cracks leading into the soil. Where runoff is not a factor, moisture conditions most favorable for plants were found in areas with the following characteristics: (i) a level or nearly level landscape that is stable and shows little or no evidence of erosion, (ii) a thin, coarse-textured horizon at the surface for maximum infiltration of moisture, and (iii) a finer textured horizon and/or an indurated horizon at favorable depths to capture the moisture and prevent its movement to greater depths where it would be unavailable for plant use.

INTRODUCTION

Soil moisture, as well as other climatic features, has been recognized as an important factor in soil genesis and classification. Certain soils occur only in specific climatic zones. The water status of soil continually affects soil properties through its influence on weathering, soil development, friability, and permeability (Slatyer, 1967). It is also an important factor in erosion of some soils (Russell, 1959). In *Soil Taxonomy* (Soil Survey Staff, 1973), the

---

[1] Cooperative investigations of the Agricultural Research Service, U. S. Department of Agriculture; Soil Survey Investigations, Soil Conservation Service, U. S. Department of Agriculture; and the Agricultural Experiment Station, New Mexico State University, Las Cruces, New Mexico. Journal Article 396, New Mexico Agricultural Experiment Station.

[2] Range Scientist, Jornada Experimental Range, Agricultural Research Service, U. S. Department of Agriculture, Las Cruces, New Mexico; and Soil Scientist, Soil Survey Investigations, Soil Conservation Service, U. S. Department of Agriculture, University Park Branch, Las Cruces, New Mexico, respectively.

system of soil classification adopted by the National Cooperative Soil Survey, soil moisture characteristics are used as differentiating criteria.

Soil moisture and precipitation data have been collected at a number of sites in an arid region of southern New Mexico. The effects of precipitation on soil moisture depend on such factors as (i) soil characteristics (e.g., structure and texture), (ii) position on the landscape, (iii) amount and intensity of precipitation event, (iv) plant cover, and (v) soil moisture status at time of storm.

There is very little information in the literature concerning field measurement of soil moisture on arid rangelands. Winkworth (1970) studied the soil water regime of an arid grassland [*Eragrostis eriopoda* Benth.] in central Australia. During a 2-year period, there were six significant periods of soil water recharge followed by withdrawal of soil moisture to an average soil water potential of −120 bars. Shreve (1934) demonstrated the relationship between the amount and intensity of rainfall and the periodicity of soil moisture near Tucson, Arizona. Kincaid, Gardner, and Schreiber (1964) compared soil moisture on a grass-covered drainage area of noncalcareous soil and on a shrub-covered drainage area of calcareous soil in southeastern Arizona. Moisture depletion at the 15-cm depth was more rapid in the grassland area indicating greater evapotranspiration. Moisture withdrawal from the 45-cm depth was about the same on the two areas. Cumulative infiltration increased with surface gravel. Houston (1968) compared seasonal soil moisture accumulation and depletion for different soils on heavily and lightly grazed semiarid rangelands in Montana.

MATERIALS AND METHODS

This study was conducted on the Jornada Experimental Range near Las Cruces, New Mexico (Fig. 1). The mountain ranges are steep and rocky. Large alluvial fans occur at the mouths of major canyons along the mountain fronts. Arroyos occur between the fans and cross them in places. The arroyos are dry, except when runoff occurs from precipitation on areas upslope. Downslope the fans merge into a broad, coalescent fan-piedmont with a slope of 2 to 3%. Slopes gradually decrease to about 0.5% at toeslopes of the fan-piedmont along the margins of the basin floor. Relief in the basin floor ranges from level to gently undulating, and there are scattered playas (see Fig. 2 for the location of the study sites and the general topography of the area). Table 1 gives the soils, landscapes, and geomorphic surfaces of the various sites.

The average annual precipitation is 22 cm; an average of 13 cm occurs during July to September. Most winter moisture comes from low-intensity rains or occasionally from snow. Most summer rainfall occurs as localized thunderstorms of high intensity. The precipitation is highly variable from time to time and place to place. Springs are usually dry and windy. The average annual evaporation from a Weather Bureau pan is 225 cm. The

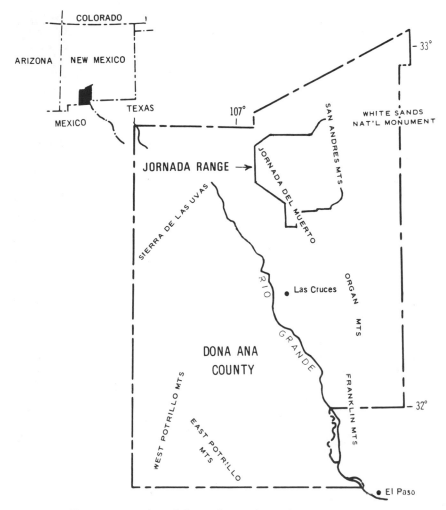

Figure 1. Location of the study area in southern New Mexico.

average temperature is 4C in January and 26C in July. The average annual precipitation for 1961–70 was 22 cm, or the same as the long time average.

Vegetation data were collected at the end of the summer growing season in close proximity to the soil moisture station. Perennial grasses were clipped at ground level; old growth was separated from the production of the current year and the old growth was discarded; and the herbage was air-dried and weighed.

Soil parent materials in the basin floor are sandy sediments of the ancestral Rio Grande, the fluvial facies of the Camp Rice Formation (Strain, 1966; El Paso Geological Society, 1970). In places, the fluvial sediments have been moved by wind. Sediments of the alluvial fan-piedmont adjacent to the basin floor were derived from the San Andres and Dona Ana Moun-

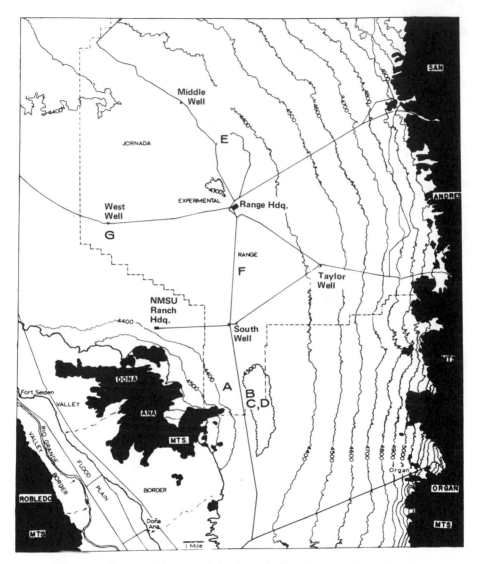

Figure 2. Topographic map of the Jornada Experimental Range showing the location of study sites A-G. Dashed lines indicate the approximate boundary of the Experimental Range.

tains in the area of the Experimental Range. Sedimentary rocks, including limestone, are dominant in the San Andres Mountains (Kottlowski et al., 1956). The Dona Ana Mountains contain a variety of igneous rocks, primarily monzonite, rhyolite, andesite, and latite, with a small area of sedimentary rocks (Kottlowski, 1960). In addition to the alluvial sediments, atmospheric additions of calcium and clay form an important part of the parent materials particularly in older soils. These atmospheric additions

Table 1. Some characteristics of soils and landscapes of the study area.

Site*	Soil series, variant or phase	Classification	Landscape position and slope	Geomorphic surface and age
A	Canutio, coarse-loamy variant	Typic Torriorthent, coarse-loamy, mixed, thermic	Fan-piedmont slope 2%	Organ, Holocene
B	Stellar, wedgy subsoil variant	Ustollic Haplargid, fine, mixed, thermic	Fan-piedmont toeslopes slope 0.5%	Jornada II, late-Pleistocene
C	Stellar	Ustollic Haplargid, fine, mixed, thermic	Basin floor nearly level	Jornada I, late mid-Pleistocene
D	Reakor†	Ustollic Calciorthid, fine-silty, mixed, thermic	Basin floor nearly level	Petts Tank, late-Pleistocene
E	Algerita†, deep gypsum phase	Typic Calciorthid, fine-loamy, mixed, thermic	Playa, level	Alluvium in floor of playa (latest Pleistocene-Holocene?)‡
F	Onite, buried soil variant	Typic Haplargid, coarse-loamy, mixed, thermic	Basin floor nearly level	Apparent eolian accumulation (Holocene?) on La Mesa surface (mid-Pleistocene)‡
G	Hueco†	Petrocalcic Paleargid, coarse-loamy, mixed, thermic	Basin floor nearly level	La Mesa, mid-Pleistocene

* Cf. Fig. 2.
† Tentative series.
‡ Details of the stratigraphy and chronology are not known in this part of the Experimental Range and assigned chronology must be considered tentative.

come from both the dry dustfall and the precipitation (Gile, Hawley, & Grossman, 1970). The top of the zone of ground-water saturation in basin-fill deposits of the study area commonly ranges from about 90 to 125 m (King et al., 1971).

Most of the Experimental Range is located north of the Desert Soil-Geomorphology Project where detailed soil-geomorphic studies have been conducted (see Gile et al., 1970 for a summary discussion and bibliography of some of this work). However, some of the conclusions from the Desert Project studies may be extrapolated to the Experimental Range.

Soil Moisture

Gypsum electrical resistance blocks were placed at several locations at varying depths depending on the soil depth. The soil moisture stations were located within a livestock exclosure. Moisture potential measurements were recorded with an ohmmeter 2 or 3 times a week when there was moisture during the summer. They were recorded monthly during the remainder of the year when there were fewer changes in moisture status. The blocks were calibrated in light- and medium-textured soils. They gave similar readings at the same moisture potential for the two soils. All of the blocks were tested and only those with similar response curves were used. The calibration and use of gypsum blocks has been discussed by Taylor, Evans, and Kemper (1961). The blocks used in this study were purchased from Taylor. Slatyer (1967) and Kramer (1969) are two of the recent authors who discuss use of gypsum-impregnated blocks to measure matric potential of soil water.

At sites C, D, and E, soil moisture was measured both inside and outside a sheet metal cylinder. The cylinder was 3 m in diameter and was buried

15 cm in the soil. The soil moisture units inside the cylinder provided estimates of moisture due to precipitation. Those outside the cylinder provided
estimates of moisture due to precipitation plus run-in.

Soil temperatures were also recorded at several depths at several locations at about the same time as moisture was recorded. All of the resistance
readings were adjusted to 15.6C. Precipitation was recorded at each study
site in a standard U. S. Weather Bureau rain gauge.

For purposes of this study, daily precipitation of less than 0.63 cm was
omitted from consideration. Examination of soil moisture data at the 10-cm
depth (and consideration of the high rate of evaporation) showed that light
amounts of precipitation seldom affected soil moisture at that depth. Bailey
found that individual precipitation events of 1.3 cm or less during the summer
were ineffective on some of the same sandy soils used in this study (O. F.
Bailey, 1967. Water availability and grass root distribution in selected soils.
M.S. Thesis. New Mexico State University, Las Cruces.).

The number of days when the moisture potential was between 0 and
−15 bars at each depth was determined for each year. This was correlated
with the annual precipitation (omitting daily amounts of less than 0.63 cm).
A regression equation was computed for each depth that had a significant
correlation (0.05 level).

Soil Morphology

The soils around the blocks were examined in detail with an auger, without
disturbance of the blocks themselves, to determine the character of the soils
in which the gypsum blocks were embedded. A pit was then dug in a similar
soil, usually at a distance of only several meters away from the actual site of
moisture measurement. The soils were described and then classified (Soil
Survey Staff, 1973). Soil texture, type of structure, and dry consistence are
summarized in Tables 2–5. Full descriptions are given in the Appendix along
with notes about classification of the soils. Although no laboratory data are
available at the actual sites of moisture measurement, a number of soils
sampled in the Desert Project are similar. This information has been considered in the classification and genetic discussion of the soils at the moisture
stations.

Two of the most prominent horizons in soils of this arid region are
horizons of silicate clay and carbonate accumulation. Smith and Buol (1968)
concluded that illuviation of silicate clay had occurred in certain semiarid
soils that they studied in Arizona. In southern New Mexico, coatings of
oriented clay on sand grains are typical of these horizons of silicate clay accumulation (Gile & Grossman, 1968). These oriented coatings are most
strongly expressed in the horizon of maximum silicate clay. The correspond-
ence of these factors is taken as evidence that oriented clay coatings on sand
grains may be used as a marker of clay illuviation. Reference is made to these
clay coatings in soil descriptions given in the Appendix. Morphogenetic rela-

tionships also indicate that carbonate in horizons of carbonate accumulation in soils of the region is predominantly to wholly of illuvial origin (Gile, Peterson, & Grossman, 1966).

Morphological features of Holocene soils are helpful in estimating common depths of present wetting because the morphology of such soils cannot be attributed to periods of more effective moisture during the Pleistocene (Gile, 1970). These morphological features may then be used in an assessment of Holocene processes of pedogenesis in soils that developed in part during a Pleistocene pluvial. Hence, the soil moisture data of this study which have been measured in soils of widely variable ages (from Holocene to mid-Pleistocene) provide quantitative information concerning amounts, depths, and times of wetting associated with the development of these morphological features.

Soil horizon terminology follows the *Supplement to the Soil Survey Manual* (Soil Survey Staff, 1962) except for the following: (i) Arabic instead of Roman numerals are prefixed to the master horizon designations to indicate lithologic discontinuities; (ii) wedge-shaped structural aggregates are designated wedges (Soil Survey Staff, 1973); (iii) horizons showing structural development in the B position are designated B; (iv) the K horizon nomenclature (Gile et al., 1965, 1970) is used to designate horizons of prominent carbonate accumulation; and (v) the C horizon is reserved for horizons that show little or no evidence of pedogenesis.

RESULTS AND DISCUSSION

Data are presented for certain soils of three general landscape positions: (i) a basin floor and an adjacent fan-piedmont that contributes water to the basin floor, (ii) a playa that receives some run-in water from long drainageways leading to adjacent mountains, and (iii) a broad, nearly level basin floor with good infiltration, only localized runoff, and no run-in from adjacent slopes.

Basin Floor, With Run-in, and Adjacent Fan-Piedmont

The landscape of this area (sites *A*, *B*, *C*, and *D*, Fig. 2) consists of a nearly level basin floor and an adjacent alluvial fan-piedmont. The study area on the fan-piedmont has 0.5-2% slopes that contribute water to the basin floor following rainfalls of high intensity. Sites *A* and *B* (Fig. 2) are on the fan-piedmont; sites *C* and *D* are on the basin floor.

SITE A

Vegetation records (1858 and 1915) and photographs (about 1920) indicate that this site was dominated at that time by black grama [*Bouteloua eriopoda*

(Torr.) Torr.] (Buffington & Herbel, 1965). The area is now dominated by creosotebush [*Larrea tridentata* (DC.) Coville]. With this shift in vegetation, there has been some erosion in the general area, and scattered small drainageways occur in the vicinity. The soil moisture units are located in a relatively stable area between the drainageways. The slope is 2%.

Examination of the soil adjacent to the moisture blocks indicates that they are embedded in a Typic Torriorthent, coarse-loamy, mixed, thermic. The texture, type of structure, and dry consistence of a similar soil (Canutio, coarse-loamy variant) near the moisture blocks are given in Table 2. The soil has formed in fan-piedmont sediments derived from monzonite, andesite, and rhyolite. This soil is on the Organ surface of Holocene age (Ruhe, 1967; Hawley & Kottlowski, 1969) and is less than 5,000 years old (Gile et al., 1970).

The average annual precipitation (excluding daily amounts of less than 0.63 cm) during 1960-70 was 20.8 cm (Table 2). The number of annual precipitation events associated with the total ranged from 7 to 25. During that period, soil moisture was recorded at depths of 10, 25, 40, 60, and 90 cm. The average number of days per year when the soil moisture potential was between 0 and −15 bars ranged from 129.9 days at the 10-cm depth to only 6.6 days at the 90-cm depth. During a wet year, 1961, there were 251 days when the moisture potential was between 0 and −15 bars at the 25-cm depth as follows: 1 January–18 May, 17–30 July, 17 August–28 September, and 20 November–26 December. At the 40-cm depth, no moisture between 0 and −15 bars was recorded for 5 of the 11 years. At the 90-cm depth, it was 9 out of the 11 years. Only the number of days with soil moisture at the 10- and 60-cm depths was significantly correlated to annual precipitation, probably because there is considerable runoff from the area.

Since this soil (to a depth of 67 cm, see Site A, Appendix) is less than 5,000 years old, it must have formed in a climate similar to the present one. The moisture data show that the noncalcareous B horizon is wetted most frequently, while the 2C1ca horizon is wetted somewhat less frequently. The moisture data, soil morphology, and soil age indicate that silicate clay is slowly accumulating in the B horizon and carbonate in the Cca horizon.

SITE B

The second site on the fan-piedmont occurs on the toeslopes at the edge of the basin floor. The slope is about 0.5%. While fan-toeslopes, such as this site, may contribute some runoff water to the basin floor from a high-intensity storm, they also receive considerable run-in from the adjacent steeper slopes such as those at site A. There are no distinct drainageways at this site but there is a microrelief, a few decimeters wide, caused by small depressions. The vegetation is a dense stand of tobosa [*Hilaria mutica* (Buckl.) Benth.]. Average annual production for 1958–61 was 3,000 kg/ha (Herbel, 1963). For 1960–66, it was 2,344 kg/ha. Clipping on this site was discontinued following 1966.

The soil moisture units are placed in an Ustollic Haplargid, fine, mixed, thermic. The texture, type of structure, and dry consistence of a similar soil (Stellar, wedgy subsoil variant) near the moisture blocks are given in Table 2. The soil has formed in fan-piedmont sediments derived from monzonite, rhyolite, andesite, and latite, and is on the Jornada II surface of late-Pleistocene age (Gile & Hawley, 1968).

The average annual precipitation during 1960-70 was 19.8 cm (Table 2). During that period, soil moisture was recorded in two replications at depths of 10, 25, 40, 60, 90, and 120 cm. Soil moisture potential between 0 and −15 bars was recorded at an annual average of 166.5 days at the 25-cm depth and 203.9 days at the 60-cm depth. There were 46 days in 1960, 7 July-2 August and 14 August-3 September, when the moisture potential

Table 2. Precipitation, soil moisture, and soil morphology at Sites A and B, on the fan-piedmont adjacent to the basin floor.

Soil moisture and precipitation			Soil morphology (in part)¶			
	Site A	Site B		Site A		Site B
Precipitation (cm)*				Typic Torriorthent, coarse-loamy, mixed, thermic		Ustollic Haplargid, fine, mixed, thermic
Mean	20.8	19.8				
Range	10.5-32.5	8.0-32.0		Horizon and depth, cm	Morphology¶	Horizon and depth, cm	Morphology¶
Soil moisture (days)† at stated depths						
10 cm				A2, 0-4	Fine sandy loam, platy, crumb, soft, loose	A2, 0-7	Loam, crumb, blocky, slightly hard
Mean	129.9	172.1				
Range	29-227	59-280				
r‡	0.61	0.52		B, 4-20	Gravelly sandy loam, massive, slightly hard	B1t, 7-15	Clay loam, blocky, very hard
Regression§	19.3+5.3X	--				
25 cm						
Mean	107.7	166.5				
Range	1-251	46-285				B21t, 15-43	Clay, blocky, extremely hard
r	0.54	0.16		2C1ca, 20-33	Very gravelly sandy loam, massive, soft		
Regression	--	--				B22t, 43-75	Clay, wedgy, platy, extremely hard
40 cm						
Mean	60.8	191.1				
Range	0-304	63-301		3C2ca, 33-52	Gravelly sandy loam, blocky, slightly hard	K & B, 75-90	Clay, blocky, very hard
r	0.56	0.21				
Regression	--	--				K2, 90-112	Silty clay loam, blocky, hard
60 cm				3C3ca 52-67	Gravelly sandy loam, massive, soft		
Mean	58.5	203.9				Bcacs, 112-133	Clay loam, blocky, hard
Range	0-224	51-356				
r	0.67	0.25		3Btcab, 67-80	Gravelly sandy clay loam, blocky, slightly hard	B1tcacsb, 133-152	Clay loam, blocky, hard
Regression	-107.6+8.0X	--				
90 cm						B2tcacsb, 152-163	Clay loam, blocky, hard
Mean	6.6	175.5				
Range	0-41	0-321		4K&Ccab, 80-102	Gravelly sandy loam, massive, slightly hard		
r	0.52	0.31				
Regression	--	--				
120 cm				5Btcab2, 102-123	Gravelly sandy clay loam, blocky, hard		
Mean	--	187.5				
Range		0-365				
r		0.26				
Regression		--				

* Annual precipitation excluding daily amounts of less than 0.63 cm.
† Days per year with the moisture potential between 0 and -15 bars.
‡ Simple correlation between annual precipitation and days of moisture at each depth.
§ Regression equations for those locations and depths having significant correlations (0.05 level).
  (Y = number of days of soil moisture at that depth and X = annual precipitation.)
¶ Dominant texture, type of structure, and dry consistence for the stated horizons and depths. See sites A and B, Appendix, for full description.

was between 0 and −15 bars at the 25-cm depth. During 1962, there was moisture at the 25-cm depth for 285 days: 1 January–27 May, 20 July–29 August, and 26 September–30 December. There was only 1 year when no moisture was recorded at the 90- and 120-cm depths. There was no significant correlation of soil moisture at any depth with precipitation at this location, probably because a variable amount of the soil moisture is attributable to run-in water.

Upper horizons of this soil are quite high in clay which should cause slow infiltration of moisture from the surface. For this reason, the depth of moisture penetration (Table 2) might seem surprising at first glance. The microrelief and soil morphology suggest an explanation for the relatively deep and frequent wetting of this soil. A network of small circular and linear depressions occurs just upslope from the moisture station. One of the linear depressions is about 5 m long, 20–30 cm deep, and 50–75 cm wide; it occurs along the contour and should catch moisture moving downslope. This linear depression has branches 3–5 m long, one of which leads directly towards the moisture blocks and ends only about 1 m away from the blocks. In places, small tubes[3] extend from the depressions into the Bt horizon. The Bt horizon itself, described during a dry time, should transmit water rapidly at the onset of rainfall or run-in because of the numerous cracks between the plates, wedges, and blocks. These relationships indicate that the depressions, the tubes in the soil surface, and the cracks in the Bt horizon could cause lower horizons to be moistened quite rapidly and to depths greater than if wetting were accomplished only by downward movement through the matrix of the overlying horizons.

Soil morphology provides additional supporting evidence that the soil is wetted deeply in this way. The Stellar, wedgy subsoil variant is strongly calcareous throughout. Adjacent Argids only a few meters away, on the same surface, of the same age, and without the depressions, have Bt horizons that are noncalcareous in their upper parts and that lack the wedges and plates. This indicates that carbonates are being quite evenly leached by downward movement of water in the adjacent Argids, but not in the Argids with plates and wedges. These relationships also indicate considerable differences in depths of moisture penetration in soils that are only a few meters apart.

SITE C

Both this site and site D occur on the relatively level basin floor, although run-in water from the adjacent slopes does not stand on the area but drains slowly to a playa about 2 km south of these sites. The vegetation on this site is also tobosa, but in a somewhat sparser stand than at the previous location. Average annual production for 1960–70 was 1,055 kg/ha; during 1960–66 it was 755 kg/ha. The latter was only 32% of the production obtained at the previous site for the same period.

---

[3]A term used to designate holes (in this case about 5 to 15 cm in diameter) descending into the soil.

The soil moisture units are placed in an Ustollic Haplargid, fine, mixed, thermic. The texture, type of structure, and dry consistence of a similar soil (Stellar) near the moisture blocks are given in Table 3a. The soil has formed in basin floor sediments derived from monozonite, rhyolite, and andesite, and is on the Jornada I surface of late mid-Pleistocene age.

The average annual precipitation during 1960–70 was 19.6 cm (Table 3b). During that period, soil moisture was recorded in two replications at depths of 10, 25, 40, 60, 90, and 120 cm at site C, both inside and outside a metal cylinder. The soil moisture potential was between 0 and −15 bars for 48–53 days at the 40- through 120-cm depths for the 1960–70 period outside the metal cylinder. It was 82 and 134 days for the 25- and 10-cm depths, respectively. At the 25-cm depth outside the cylinder, no moisture between 0 and −15 bars potential was recorded for 2 of the 11 years. There was moisture for 205 days at the 25-cm depth in 1962: 1 January–8 May, 20 July–12 August, and 28 September–24 November. No moisture was recorded for 7 of the 11 years at the 90- and 120-cm depths. Inside the cylinder, the soil moisture potential was between 0 and −15 bars for 92 days at the 10-cm depth and for 32–40 days for the remaining depths. At the 25-cm depth inside the cylinder, no moisture between 0 and −15 bars potential was recorded for 7 of the 11 years. A similar situation existed for all depths greater than 25 cm.

Comparing data inside and outside the cylinder, an average of 36% of the days with soil moisture was attributable to run-in. Comparing data from site B and site C (outside the cylinder), there was an average of 128–425% more days with soil moisture potential between 0 and −15 bars at site B for the various depths. Outside the cylinder at site C, the number of days with soil moisture at the 10- through 60-cm depths was significantly correlated with annual precipitation. Inside the cylinder, all correlation values were significant and higher than those obtained from soil moisture measurements outside the cylinder.

The soil surface, morphology, and moisture data indicate that present genetic processes of this soil differ from those of the soil at site B. The Haplargid at site B is moistened deeply by way of the depressions and small tubes leading to cracks in the Bt horizon. In contrast, the depressions and tubes are not present at site C, and moistening of this soil is accomplished by wetting through the soil matrix from the soil surface downward. The soil moisture decreases regularly from the top down, in contrast to the situation at site B. The regular occurrence of the noncalcareous A3 and B21t horizons and the underlying calcareous B22t horizon in the soil at site C reflects this regular pattern of soil moisture. The moisture data and morphology of this soil and of the soil of Holocene age at site A, together with morphological comparisons of soils of Holocene and Pleistocene age elsewhere in the Desert Project (Gile, 1970), indicate that silicate clay is probably slowly accumulating in the B21t horizon at the present time. Carbonate accumulation is restricted primarily to the middle and lower parts of the B horizon with lesser accumulation at underlying depths.

The calcareous state of the A2 horizon is a feature of many soils in this basin-floor position. The carbonate has apparently been emplaced by the moisture of carbonate-laden waters and by incorporation of carbonate in the dustfall. More carbonate has accumulated at the soil surface than can be moved down into the soil by present soil moisture.

The effect of run-in is substantial (Table 3). At the 25-cm depth, the days per year when soil moisture is at 0 to −15 bars with run-in is more than twice that without run-in. This extra moisture may be critically important in keeping the A3 and B21t horizons noncalcareous.

However, available soil moisture at all depths, both with and without run-in, is less frequent than for the Paleargid at site G to be discussed later. This is attributed to the considerably greater percentage of clay and silt associated with the smooth-topped plates in the A2 horizon. Such horizons have a tendency to "seal" when wetted (Gile et al., 1970).

SITE D

This site is about 75 m from site C. The major species on this site is burro-grass [*Scleropogon brevifolius* Phil.]. Average annual production for 1960–70 was 651 kg/ha or about 62% of the production obtained at site C.

The soil moisture blocks are in a Ustollic Calciorthid, fine-silty, mixed, thermic. The texture, type of structure, and dry consistence of a similar soil

Table 3a. Soil morphology at Sites C and D in the basin floor.

Soil morphology (in part)*			
Site C		Site D	
Ustollic Haplargid, fine, mixed, thermic		Ustollic Calciorthid, fine-silty, mixed, thermic	
Horizon and depth, cm	Morphology*	Horizon and depth, cm	Morphology*
A2, 0–5	Clay loam, platy, slightly hard	A, 0–4	Clay loam, platy, slightly hard
A3, 5–9	Clay loam, blocky, hard	A3, 4–10	Silty clay loam, blocky, very hard
B21t, 9–23	Clay, prismatic, blocky, very hard	B1, 10–26	Clay loam, blocky, slightly hard
B22t, 23–44	Clay loam, prismatic, blocky, very hard	B21ca, 26–53	Silty clay loam, prismatic, blocky, hard
B23tca, 44–67	Clay, prismatic, blocky, very hard	B22ca, 53–79	Clay loam, prismatic, blocky, hard
B24tca, 67–87	Clay, prismatic, blocky, very hard	K & B, 79–95	Clay loam, blocky, slightly hard
K & Bt, 87–118	Clay loam, blocky, hard	B2tcab, 95–122	Sandy clay loam, prismatic, blocky, hard
K21, 118–134	Clay loam, platy, hard		

* Dominant texture, type of structure, and dry consistence for the stated horizons and depths. See sites C and D, Appendix, for full descriptions.

(Reakor) near the moisture blocks are given in Table 3a. The soil has formed in basin floor sediments derived primarily from sedimentary rocks such as limestone, siltstone, sandstone, and shale with lesser amounts derived from igneous rocks. This soil occurs on the Petts Tank surface (Hawley & Gile, 1966).

During 1960–70, soil moisture was recorded in two replications at the same depths as sites B and C, both inside and outside a sheet metal cylinder. Soil moisture potential between 0 and −15 bars ranged from 114.9 days at

Table 3b. Precipitation and soil moisture at sites C and D in the basin floor.

	Soil moisture and precipitation			
	Site C*	Site C[†]	Site D*	Site D[†]
Precipitation (cm)[‡]				
Mean	19.6	19.6	19.6	19.6
Range	11.8–32.0	11.8–32.0	11.8–32.0	11.8–32.0
Soil moisture (days)[§] at stated depths				
10 cm				
Mean	134.4	91.8	114.9	78.2
Range	33–205	18–172	46–192	0–175
r[¶]	0.65	0.79	0.88	0.78
Regression**	23.8+5.7X	−29.6+6.2X	−7.9+6.3X	−49.6+6.5X
25 cm				
Mean	82.4	40.4	69.6	45.3
Range	0–205	0–179	0–165	0–162
r	0.80	0.83	0.82	0.87
Regression	−103.7+9.5X	−114.5+7.9X	−81.8+7.7X	−113.4+8.1X
40 cm				
Mean	51.3	32.7	47.2	28.3
Range	0–185	0–186	0–172	0–173
r	0.71	0.72	0.67	0.70
Regression	−74.5+6.4X	−100.4+6.8X	−69.1+5.9X	−96.1+6.4X
60 cm				
Mean	48.0	33.1	49.7	29.0
Range	0–193	0–182	0–181	0–174
r	0.66	0.72	0.51	0.70
Regression	−83.0+6.7X	−99.3+6.8X	--	−96.1+6.4X
90 cm				
Mean	52.6	32.8	34.5	27.3
Range	0–196	0–181	0–181	0–155
r	0.50	0.71	0.68	0.69
Regression	--	99.0+6.7X	−93.4+6.5X	−96.1+6.4X
120 cm				
Mean	51.7	37.5	38.0	27.3
Range	0–176	0–203	0–181	0–155
r	0.42	0.66	0.67	0.69
Regression	--	−91.6+6.6X	−91.9+6.6X	−96.1+6.4X

* Soil moisture due to precipitation plus run-in.
[†] Soil moisture due to precipitation only.
[‡] Annual precipitation excluding daily amounts of less than 0.63 cm. Because of the proximity of sites C and D, precipitation was recorded only at site C.
[§] Days per year with the moisture potential between 0 and −15 bars.
[¶] Simple correlation between annual precipitation and days of moisture at each depth.
** Regression equations for those locations and depths having significant correlations (0.05 level).

the 10-cm depth at site D to 34.5 days at the 90-cm depth outside the cylinder (Table 3b). At the 25-cm depth outside the cylinder, no moisture between 0 and −15 bars potential was recorded for 2 of the 11 years. During the wettest year, 1967, at the 25-cm depth, moisture was recorded 3–29 July and 14 August–31 December. No moisture was recorded for 8 of the 11 years at the 90- and 120-cm depths. Inside the cylinder, the soil moisture potential was between 0 and −15 bars for 78.2 days at the 10-cm depth and for 27.3 days at the 90- and 120-cm depths. There was 1 year when no moisture between 0 and −15 bars was recorded at the 10-cm depth. It was 6 years at the 25-cm depth and 9 years for all the remaining depths. Over all sampling depths, there were about 33% more days with moisture between 0 and −15 bars outside the cylinder or due to run-in. Virtually all correlations between days with soil moisture and precipitation were significant.

This soil has formed in high-carbonate parent materials as noted earlier. Studies of a similar soil not far from this site indicate that some carbonate originally in the parent materials still is present in upper horizons (Gile et al., 1970). Present moisture is clearly insufficient to remove carbonate from upper horizons. Further, this soil was formed in part during a Pleistocene pluvial when there was more effective moisture than now. The relationships indicate that even greater amounts of effective moisture of a pluvial was insufficient to remove the carbonate from upper horizons. As with the Haplargid at site C, the increased days of moisture due to run-in are considerable.

Playa

SITE E

This site is in a small playa at the end of a drainageway that begins on the slopes of the San Andres Mountains. The playa is flooded about twice every 3 years. The vegetation on this site is a mixed stand of tobosa and burrograss. Separate yield estimates were obtained for the relatively unmixed patches of each species. Average annual production of tobosa for 1960–70 was 1,291 kg/ha. Burrograss yields for the same period were 637 kg/ha. The yields at this site are more variable than those reported for sites B, C, and D.

The soil moisture blocks are embedded in a Typic Calciorthid, fine-loamy, mixed, thermic. The texture, type of structure, and dry consistence of a similar soil (Algerita, deep gypsum phase) near the moisture blocks are given in Table 4. The soil has apparently formed primarily in alluvium resting on gypsum of lacustrine origin. While the alluvium has not been studied in detail, it apparently was originally derived from rocks of the San Andres Mountains to the east—limestone, sandstone, siltstone, and shale, with lesser amounts of igneous rocks. The alluvium is thickest in the lowest part of the playa, as at this site. At other parts of the playa, gypsum is nearer the surface or at the surface.

Table 4. Precipitation, soil moisture, and soil morphology at Site E, in the playa.

Soil moisture and precipitation			Soil morphology (in part)*	
	Site E[†]	Site E[‡]	Site E	
Precipitation (cm)[§]			Typic Calciorthid, fine-loamy, mixed, thermic	
Mean	18.8	18.8		
Range	8.9–31.1	8.9–31.1	Horizon and depth, cm	Morphology*
Soil moisture (days)[¶] at stated depth				
10 cm			C, 0–3	Fine sandy loam, single grain, loose
Mean	124.0	112.7		
Range	32–211	36–176	A1, 3–12	Fine sandy loam, massive, platy, hard
r**	0.50	0.45		
Regression[††]	--	--		
25 cm			B11, 12–31	Clay loam, prismatic, blocky, very hard
Mean	81.2	71.5		
Range	0–227	0–190		
r	0.46	0.46	B12, 31–50	Sandy clay loam, prismatic, blocky, very hard
Regression	--	--		
40 cm				
Mean	45.6	38.1	B21ca, 50–65	Silty clay loam, prismatic, blocky, slightly hard
Range	0–168	0–146		
r	0.50	0.62		
Regression	--	-40.2+4.2X	B22ca, 65–90	Silty clay loam, prismatic, blocky, slightly hard
60 cm				
Mean	6.8	1.9		
Range	0–30	0–21	B23ca, 90–112	Silty clay loam, prismatic, blocky, hard
r	0.72	-0.21		
Regression	-14.5+1.1X	--		
90 cm			2C1ca, 112–120	Loam, prismatic, blocky, slightly hard
Mean	0	0		
Range				
r				
Regression			2C2ca, 120–138	Sandy loam, massive, very hard
120 cm				
Mean	0	0		
Range				
r				
Regression				

* Dominant texture, type of structure, and dry consistence for the stated horizons and depths. See Site E, Appendix, for full description.
† Soil moisture due to precipitation plus run-in.
‡ Soil moisture due to precipitation only.
§ Annual precipitation excluding daily amounts of less than 0.63 cm.
¶ Days per year with the moisture potential between 0 and -15 bars.
** Simple correlation between annual precipitation and days of moisture at each depth.
†† Regression equations for those locations and depths having significant correlations (0.05 level). (Y = number of days of soil moisture at that depth, and X = annual precipitation.)

The average annual precipitation during 1960–70 was 18.8 cm (Table 4). During that period, soil moisture was recorded on an area with a mixed stand of tobosa and burrograss in two replications at depths of 10, 25, 40, 60, 90, and 120 cm, both inside and outside a sheet metal cylinder. During the 11-year period, no moisture between 0 and -15 bars potential was recorded at the 90- and 120-cm depths. In 1959, there were 31 days with moisture between those levels at those depths. Soil moisture was recorded for an average of 124.0 days at the 10-cm depth and 6.8 days at the 60-cm depth outside the cylinder. At the 25-cm depth outside the cylinder, no moisture between 0 and −15 bars potential was recorded for 4 of the 11 years. There was moisture for 227 days at the 25-cm depth in 1962: 1 January–29 April, 28 July–17 August, 3 October–3 December, and 29–31 December. No moisture was recorded for 8 of 11 years at the 60-cm depth. Inside the cylinder, the soil moisture potential was between 0 and −15 bars for 112.7 days at the 10-cm depth and 1.9 days at the 60-cm depth.

Over all depths with moisture, there was an average of about 8 days each year when soil moisture could be attributed to run-in. This is considerably lower than for sites B, C, and D. This is attributed to the fact that the local watershed is much less than at sites B, C, and D. Also, observations indicate that water from the mountains and intervening areas does not often reach the playa in large amounts. The days with soil moisture at only one depth in each, outside and inside the cylinder, were significantly related to precipitation.

The moisture data show that the A1 and B11 horizons are moistened most frequently. However, these horizons are still strongly calcareous. This is probably due in part to the highly calcareous nature of the parent materials.

Judging from morphology and soil moisture relations at sites A, C, and D, carbonate is slowly accumulating at the present time in the soil at site E, primarily in the upper part of the B horizon. No evidence of illuviation of silicate clay was observed in the soil at site E.

Basin Floor Without Run-in

This part of the basin has a level to gently rolling topography. Water movement following high intensity rainfall is very localized because the landscape is level or nearly level; there is no run-in; and upper horizons are relatively coarse in texture.

SITE F

This site is on a very slight slope. The area is subject to wind erosion and the surface has small hummocks. This area had a good cover of black grama prior to the great drouth of 1951–56 (Herbel, Ares, & Wright, 1972). By the end of that drouth, black grama cover was reduced to 1% of the predrouth average and it has not increased since that time. The present vegetation is a relatively sparse stand of mesa dropseed [*Sporobolus flexuosus* (Thurb.) Rydb.] and a variety of forbs and annual grasses under certain weather conditions. Average annual production of perennial grasses for 1960–70 was 23 kg/ha.

Table 5. Precipitation, soil moisture, and soil morphology at Sites F and G, in the basin floor.

Soil moisture and precipitation		
	Site F	Site G†
Precipitation (cm)‡		
Mean	19.3	18.9
Range	10.8-29.8	11.5-34.9
Soil moisture (days)§ at stated depths		
10 cm		
Mean	192.0	193.9
Range	97-301	64-321
r¶	0.51	0.51
Regression**	--	--
25 cm		
Mean	173.5	212.2
Range	62-318	99-336
r	0.58	0.64
Regression	--	81.4+6.9X
40 cm		
Mean	121.9	158.5
Range	0-312	32-333
r	0.63	0.67
Regression	-80.2+10.5X	-8.2+8.8X
60 cm		
Mean	90.2	116.8
Range	0-319	0-350
r	0.44	0.70
Regression	--	-107.7+11.9X
90 cm		
Mean	18.7	96.5
Range	0-179	0-278
r	0.32	0.71
Regression	--	-114.9+11.2X

Soil morphology (in part)*			
Site F		Site G	
Typic Haplargid, coarse-loamy, mixed, thermic		Petrocalcic Paleargid, coarse-loamy, mixed, thermic	
Horizon and depth, cm	Morphology*	Horizon and depth, cm	Morphology*
B2t, 0-18	Fine sandy loam, blocky, slightly hard	C, 0-5	Sand, loose, single grain, massive
B31t, 18-34	Fine sandy loam, massive, slightly hard	A2, 5-10	Fine sandy loam, massive, soft
B32t, 34-44	Loamy sand, massive, soft	B1t, 10-23	Fine sandy loam, massive, slightly hard
B1tcab, 44-60	Sandy loam, prismatic, blocky, hard	B21t, 23-36	Fine sandy loam, massive, slightly hard
B21tcab, 60-76	Sandy clay loam, prismatic, blocky, hard	B22tca, 36-46	Fine sandy loam, massive, slightly hard
B22tcab, 76-90	Sandy clay loam, prismatic, blocky, hard	B3ca, 46-71	Sandy loam, blocky, slightly hard
B23tcab, 90-103	Sandy clay loam, prismatic, blocky, hard	K1, 71-79	Very gravelly sandy loam, crumb, loose
K2b, 103-126	Sandy clay loam, blocky, very hard	K2m, 79-90	Carbonate-cemented material, massive, extremely hard

* Dominant texture, type of structure, and dry consistence for the stated horizons and depths. See Sites F and G, Appendix, for full descriptions.
† On site G, soil moisture units were placed at depths of 10, 25, 40, 53, and 68 cm instead of 10, 25, 40, 60, and 90 cm.
‡ Annual precipitation excluding daily amounts of less than 0.63 cm.
§ Days per year with the moisture potential between 0 and -15 bars.
¶ Simple correlation between annual precipitation and days of moisture at each depth.
** Regression equations for those locations and depths having significant correlations (0.05 level). (Y = number of days of soil moisture at that depth and X = annual precipitation.)

The soil moisture blocks are in a Typic Haplargid, coarse-loamy, mixed, thermic. The texture, type of structure, and dry consistence of a similar soil (Onite, buried soil variant) are given in Table 5. The soil at the land surface has apparently formed in a sandy eolian deposit considerably younger than the buried horizon below.

The average annual precipitation during 1960-70 was 19.3 cm (Table 5). During that period, soil moisture was recorded in three replications at depths of 10, 25, 40, 60, and 90 cm. The average number of days per year when the soil moisture potential was between 0 and −15 bars ranged from 192.0 at the 10-cm depth to 18.7 at the 90-cm depth. There were 62 days in 1967 (4–31 July, 21–25 August, and 29 September–30 October) when there was moisture at the 25-cm depth. During 1962, there was moisture for 318 days: 1 January–25 June, 4 July–20 August, and 28 September–30 December. At the 40-cm depth, no moisture between 0 and −15 bars was recorded for 3 of the 11 years. At the 90-cm depth, it was 8 of the 11 years. Only the number of days with soil moisture at the 40-cm depth was significantly correlated to annual precipitation. It was noted that moisture from individual precipitation events was more diffuse throughout the profile than in Site G. This has apparently contributed to greater drouth damage on this site than the shallower site G (Herbel et al., 1972).

The soil moisture at this site is considerably greater than in the soil at site E despite the fact that this soil receives virtually no run-in. This is because of the considerable difference in texture, upper horizons of this soil being much less clayey and having greater infiltration rates. This soil is non-calcareous to a depth of 44 cm and is moistened fairly frequently to that depth.

The soil at the land surface has apparently been recently truncated since the Bt horizon is at the surface. However, morphology, moisture data, and comparisons with similar soils of known age indicate that silicate clay is very slowly accumulating in the B2t and B3t horizons, and that carbonate is accumulating in the B1tcab horizon.

SITE G

There is less wind erosion evident at this site than at site F. Both areas are nearly level. There was also less drouth damage at this site than at site F. The major species on the area is black grama with scattered plants of mesa dropseed and soaptree yucca [*Yucca elata* Engelm.]. Average annual production of perennial grasses for 1960–70 was 346 kg/ha.

The soil moisture blocks are in a Petrocalcic Paleargid, coarse-loamy, mixed, thermic. The texture, type of structure, and dry consistence of a similar soil (Hueco) near the moisture blocks are given in Table 5. The soils have formed in sandy sediments of the Camp Rice Formation (fluvial facies).

The average annual precipitation during 1961–70 was 18.9 cm (Table 5). During that period, soil moisture was recorded at depths of 10, 25, 40, 53, and 68 cm. The average number of days when the soil moisture potential was between 0 and −15 bars ranged from 212.2 at the 25-cm depth to 96.5 at the 68-cm depth. During the driest year, 1965, there was moisture at the 25-cm depth 1 March–27 April and 20 September–1 November. There was moisture at the 25-cm depth for the entire year in 1961, except 11 June–10 July. At the 53-cm depth, no moisture between 0 and −15 bars potential was recorded for 3 of the 10 years. At the 68-cm depth, it was 5 of the 10 years. There was considerably more moisture at this site than at site F particularly at the deeper depth.

The soil moisture data are similar to those at site F in upper horizons, but the soil moisture increases considerably in lower horizons. The difference is partly attributed to the greater thicknesses of coarser-textured horizons that occur above the clay maximum in this soil. Infiltration rate should be relatively rapid in the uppermost two horizons, a sand and light fine sandy loam.

Another prominent difference of this soil, as compared to the one at site F, is the presence of the petrocalcic horizon at 68 cm. In general, any discontinuity, whether a change from more permeable to less permeable material with depth or the existence of relatively impermeable material within the profile, increases the water retained by the soil. At site G, the petrocalcic horizon at about the 68-cm depth holds the soil moisture at depths favorable for plant use. Further, the soil water is stored longer because the texture of the surface soil is fairly coarse. There is less loss of soil water through evaporation from soils with sandy surfaces than from soils with finer textures at the surface because of considerably reduced capillary movement of water in the coarser-textured horizons.

A large pipe (cf. Gile et al., 1966, for diagram and discussion of one of the pipes in these ancient soils of La Mesa surface) was found near the moisture blocks with an auger. The pipe extends to depths greater than 125 cm. The soil in the pipe has a fine sandy loam Bt horizon at about the same depth as the described soil. The grama grass was well developed; there appeared to be little or no difference in the vegetation of soils in the pipe and soils adjacent to the pipe.

Bailey studied infiltration and water movement on the Jornada Experimental Range on soils similar to those at sites F and G (O. F. Bailey. 1967. Water availability and grass root distribution in selected soils. M.S. Thesis. New Mexico State Univ., Las Cruces). He simulated rainfall in amounts of 1.3, 2.5, and 3.8 cm. Most of the applied water was retained in the upper two horizons of the soils studied. There was less moisture in the upper two horizons on the deeper soils, such as at site F, than the shallower soils, such as at site G, after each of the simulated rains. Apparently, the soil water is more diffused throughout the profile in the soil at site F than the shallower soil at site G.

Relation of the Soil Moisture Data to the New U. S. System of Soil Classification

The moisture data presented above are of interest from the standpoint of soil classification because criteria involving soil moisture are used in the new U. S. system of classification (Soil Survey Staff, 1973). Moreover, the data are from an arid region where very little information of this kind is available. In the classification system, *soil moisture regimes* are defined in terms of a *soil moisture control section* as discussed in the following sections.

SOIL MOISTURE CONTROL SECTION AND THE SOIL MOISTURE REGIME

The following information concerning the soil moisture criteria is quoted below since it is still in the process of publication (Soil Survey Staff, 1973).

The intent in defining the soil moisture control section is to facilitate the estimation of soil moisture regimes from climatic data. The upper boundary of this control section is the depth to which the dry(tension > 15 bars, but not air dry) soil will be moistened by 2.5 cm (1 inch) of water within 24 hours. Its lower boundary is the depth to which the dry soil will be moistened by 7.5 cm (3 inches) of water within 48 hours. These depths are exclusive of the depth of moistening along any cracks or animal burrows that are open to the surface.

If 7.5 cm of water moistens the soil to a lithic, petroferric, or paralithic contact or to a petrocalcic horizon or a duripan, the upper boundary of the rock or of the cemented horizon is the lower boundary of the soil moisture control section. If 2.5 cm of water moistens the soil to one of these contacts or horizons, the soil moisture control section is the lithic contact itself, the paralithic contact, or the upper boundary of the cemented horizon. The control section of the latter soil is moist if the upper boundary of the rock or the cemented horizon has a thin film of water. If the upper boundary is dry, the control section is dry.

As a rough guide to the limits, the soil moisture control section lies approximately between 10 and 30 cm (4 and 12 inches) if the particle-size class is fine-loamy, coarse-silty, fine-silty, or clayey. . . .The control section extends approximately from a depth of 20 cm to a depth of 60 cm (8 to 24 inches) if the particle-size class is coarse-loamy, and from 30 to 90 cm (12 to 35 inches) if the particle-size class is sandy. Obviously, coarse fragments deepen these limits to the extent that the fragments do not absorb and release water. In addition to the particle-size class, differences in structure, differences in pore size distribution, and other factors that influence movement and retention of water in the soil also affect the limits of the soil moisture control section.

The soil moisture regime, as the term is used here, refers to the presence or absence either of groundwater or of water held at a tension of less than 15 bars in the soil or in specific horizons by periods of the year. Water held at a tension of 15 bars or more is not available to keep most mesophytic plants alive. The availability of water also is affected by dissolved salts. A soil may be saturated with water that is too salty to be available for most plants, but it would seem better to call such a soil salty rather than dry. Consequently, we consider a horizon to be dry when the moisture tension is 15 bars or more. If water is held at a tension of less than 15 bars but more than zero, we consider the horizon to be moist.

Aridic and Torric (L. *aridus*, dry, and L. *torridus,* hot and dry) moisture regimes. These terms are used for the same moisture regime, but in different categories of the taxonomy.

In the aridic (torric) moisture regime, the moisture control section in most years is
a. Dry in all parts more than half the time (cumulative) that the soil temperature at a depth of 50 cm is above $5°C$; and

b. Never moist in some or all parts for as long as 90 consecutive days when the soil temperature at a depth of 50 cm is above 8°C.

The area of this study has been placed within the aridic and torric moisture regimes defined above.

The moisture data in Tables 2–5 indicate that most of the soils easily meet the requirement "In most years there is no available water in any part of the moisture control section more than half the time (cumulative) that the soil temperature at 50 cm is above 5°C. . . ."

It appears that the Haplargid at site B may not meet the stated requirements, but this soil is excluded because depths cited in the definition of the moisture control section "are exclusive of the depth of moistening along any cracks or animal burrows that are open to the surface."

The Petrocalcic Paleargid at site G does exceed the time of moistening allowed in the definition. An examination of the soil moisture records from 1961–70 indicates that available moisture was present at 25 cm more than half the time in 7 of the 10 years. However, at 40 cm (Table 5) the moisture data meet the criteria for the aridic moisture regime.

Soil moisture criteria are also used at the subgroup level of classification for some soils. Again considering the Petrocalcic Paleargid at site G, the following information (Soil Survey Staff, 1973) that applies to Petrocalcic Paleargids is pertinent:

f. are dry in all parts of the moisture control section more than three-fourths of the time (cumulative) that the soil temperature at 50-cm depth is 5°C or more.

As can be seen, the moisture control section of this soil, on an average basis, is moistened nearly three times as long as allowed under the stated definition. In the driest year, 1965, there was soil moisture between 0 and −15 bars potential for 99 days at 25 cm (Table 5). As discussed earlier, this site has no run-in but does have conditions for maximum infiltration and entrapment of moisture.

Another reason for the presence of soil moisture for a relatively long time in this soil is the precipitation that falls during some winters. Soil moisture from this precipitation tends to stay in the soil for a long time, commonly throughout the early spring season until plant growth begins.

SOIL MOISTURE AND THE USTOLLIC INTERGRADES

Aridisols in southern New Mexico are dominant in arid areas between the mountains, and Mollisols occur in places along the mountain fronts. Therefore, intergrades between these two soil orders would be expected. Such intergrades do occur and are designated as Ustollic subgroups (e.g., Ustollic Calciorthid). The identification of the Ustollic subgroups is based on the amount of organic carbon relative to sand-clay ratios (Soil Survey Staff, 1973). Ustollic intergrades occur both along the mountain fronts, closely associated with the Mollisols, and in the arid regions between the mountains.

Distribution of the Ustollic subgroups is presently under study. Indications are that they are quite readily predicted, particularly in soils that contain little or no gravel, such as the soils in this study.

Some of the arid-land soils between the mountain ranges (Fig. 2) actually contain more organic carbon than some Mollisols, but are not Mollisols because their upper horizons are too light-colored for a mollic epipedon. These soils are most common in landscape positions that receive extra moisture as runoff from areas upslope—in basin floors and broad drainageways. Also, in this area the Ustollic subgroups occur only in soils that contain fairly high amounts of clay, silt, or both. The Ustollic subgroups are well illustrated by the Stellar and Reakor soils (sites B, C, and D). The soil moisture data and analyses for similar soils indicate that some soils (e.g., the Paleargid at site G) are wetted fairly frequently in the moisture control section, but contain insufficient organic carbon because of their low clay content.

SOIL GROUPINGS BASED ON MOISTURE DIFFERENCES
WITHIN THE ARIDISOLS

As noted earlier, there is considerable variation in soil moisture in the study area even in soils only several meters apart. Because of this variation, it may be feasible to make groupings at the series or phase level where needed for use of the soil. Some separations may be made on the basis of soil morphology, such as structural differences between the Haplargids at sites B and C. Soils similar in morphology and not otherwise separable may be separated on the basis of location in landscape positions subject to overflow (an overflow phase). On a regional basis, it may be feasible to separate some soils by differences in precipitation if other factors cannot be used.

CONCLUSIONS

Much of the precipitation in this arid region falls during torrential rainstorms and runoff rates are high on slopes. Soil texture, soil structure, landscape position, vegetative cover, and microrelief are also variable and can have a marked effect on soil moisture.

Soil texture is an important factor because surficial horizons of sandy texture should have most rapid infiltration rates during rainfalls of high intensity. This is shown by the data at site G. Conversely, low infiltration rates would be expected in soil horizons with high clay content, where wetting is accomplished by infiltration at the surface and not through tubes and cracks in the soil such as at site B. At the 25-cm depth, the fine Haplargid at site C (wetted by precipitation only) had less than one-fifth the days of moisture than the coarse-loamy Paleargid at site G. Platy structure of the A horizon is also important, particularly in the finer-textured soils as such horizons tend to "seal" when wetted.

Landscape position is also significant in determining the amount of moisture that enters the soil. Runoff from higher areas can markedly increase depths of wetting in topographic lows as shown by sites C, D, and E. Conversely, such runoff decreases moisture in the soils upslope.

Microrelief can greatly influence moisture infiltration. The microrelief at site B, together with the soil morphology, indicates that the depth of moisture infiltration has been considerably increased by small depressions and tubes in the soil surface and by cracks in soil horizons connected to the tubes. Conversely, absence of these features can greatly decrease depths of wetting in soils only a few meters away.

In stable areas where runoff is not a factor, most favorable moisture conditions occur in soils and landscapes that have the following characteristics: (i) level or nearly level areas of a stable landscape that show little or no evidence of erosion, (ii) a thin surficial horizon about 5–10 cm thick and coarse textured (sand or loamy sand) to allow maximum infiltration of moisture, and (iii) a slightly finer-textured horizon (e.g., medium or heavy sandy loam) a few decimeters thick just beneath the surficial horizon to capture most of the moisture that has penetrated and to prevent its movement to greater depths with consequent loss to the plant. Such horizons still have fairly rapid infiltration rates. Indurated horizons can also be helpful in slowing or preventing downward penetration of moisture.

Therefore, precipitation values alone can be misleading from the standpoint of the moisture that actually enters the soil. Clearly the factors discussed above must be considered in evaluating moisture of arid-land soils.

APPENDIX—Descriptions of Soils at Sites A–G

Site A

Soil Variant—Canutio, coarse-loamy variant.
Classification—Typic Torriorthent, coarse-loamy, mixed, thermic.
Soil Surface—About 50% covered with rhyolite and monzonite pebbles, most of which range from 1–2 cm in diameter, with a few up to 3 cm diameter.
A2 (0–4 cm)—Pinkish-gray (7.5YR 6/3, dry) or dark brown (7.5YR 4/3, moist) fine sandy loam, mainly weak fine platy structure and soft, with some parts a loose mass of soft fine crumbs; very few roots; noncalcareous; mildly alkaline; abrupt smooth boundary.
B (4–20 cm)—Light brown (7.5YR 5.5/4, dry) or dark brown (7.5YR 4/4, moist) gravelly sandy loam; massive, slightly hard, friable; very few roots; clay coatings on sand grains and pebbles; occasional very gravelly lenses up to about 5 cm thick and 40 cm long; effervesces weakly in the lower 2 cm of the horizon where scattered pebbles have carbonate coatings on their undersides, otherwise, noncalcareous; mildly alkaline; clear wavy boundary.
2C1ca (20–33 cm)—Light brown (7.5YR 6/4, dry) or brown (7.5YR 4.5/4,

moist) very gravelly sandy loam; massive; soft, very friable; very few roots; thin carbonate coatings on pebbles; effervesces strongly; moderately alkaline; clear wavy boundary.

3C2ca (33-52 cm)—Light brown (7.5YR 6/4, dry) or brown (7.5YR 4.5/4, moist) gravelly sandy loam; weak medium subangular blocky structure; slightly hard, friable; very few roots; a few carbonate filaments; pebbles and sand grains thinly carbonate-coated; effervesces strongly; moderately alkaline; clear wavy boundary.

3C3ca (52-67 cm)—Light brown (7.5YR 6/4, dry) or brown (7.5YR 4.5/4, moist) gravelly light sandy loam; massive; soft, very friable; very few roots; thin, discontinuous carbonate coatings on pebbles and sand grains; several pockets and lenses of very gravelly materials ranging from 5-25 cm in diameter; in places the lower 1-2 cm is a fine gravelly loamy sand; effervesces strongly; moderately alkaline; abrupt and clear wavy boundary.

3Btcab (67-80 cm)—Reddish-brown (5YR 5/4, dry; 5YR 4/4, moist) gravelly light sandy clay loam; weak medium and fine subangular blocky structure; slightly hard, friable, very few fine roots; clay coatings on sand grains and pebbles; common carbonate filaments; pebbles thinly coated with carbonate; generally effervesces strongly, a few parts effervesce weakly; moderately alkaline; abrupt wavy boundary.

4K & Ccab (80-102 cm)—Dominantly pinkish-white (7.5YR 8/2, dry) or pink (7.5YR 8/4, moist) with lesser amounts of light brown (7.5YR 6/4, dry) or brown (7.5YR 5/4, moist) gravelly to very gravelly sandy loam; massive; slightly hard and hard, friable, and firm; no roots; grains in light-colored volumes are separated by carbonate; grains in darker-colored material are usually in contact but do have a few thin discontinuous carbonate coatings; in places, texture is a gravelly sand in the lower 2 cm; effervesces strongly; moderately alkaline; abrupt smooth boundary.

5Btcab2 (102-123 cm)—Reddish-brown (5YR 5/4, dry; 5YR 4/4, moist) gravelly sandy clay loam; weak medium and fine subangular blocky structure; hard, friable, and firm; no roots; clay coatings on sand grains; common carbonate filaments and pebble coatings; effervesces strongly in most places, a few areas effervesce weakly or are noncalcareous; moderately alkaline; abrupt smooth boundary to underlying petrocalcic horizon.

Remarks—The increase in clay from A to B is too slight for an argillic horizon. The B horizon is too thin for a cambic horizon since the carbonate maximum starts at 20 cm (Soil Survey Staff, 1973). There is not enough carbonate for a calcic horizon. The soil may be a Torrifluvent, but analyses of similar horizons in similar positions indicate that both amounts and irregular decrease in organic carbon are questionable for a Fluvent. Therefore, the soil is considered to be a Torriorthent.

The soil from 0 to 67 cm has formed in Organ alluvium and is less than 5000 years old. The buried soil from 67 to 102 cm is considered to be of Isaacks' Ranch age (Gile et al., 1970) and is thought to have formed largely during latest Pleistocene. The second buried soil, below 102 cm and including the petrocalcic horizon, is considered to be of Jornada II age (Gile & Hawley, 1968) and must have formed entirely during late Pleistocene.

Site B

Soil Series—Stellar, wedgy subsoil variant.

Classification—Ustollic Haplargid, fine, mixed, thermic.

Soil Surface—There are scattered small depressions, commonly 20-30 cm deep and 10-40 cm in diameter, but in places linear and ranging up to several meters long. Tubes extend from some of the depressions into the Bt horizon.

A2 (0-7 cm)—Light gray (10YR 7/2, dry) or grayish-brown (10YR 4.5/2, moist) loam; weak fine and very fine crumb in upper part, weak medium subangular blocky and coarse platy below; slightly hard and hard, friable; common roots; effervesces strongly; moderately alkaline; abrupt smooth boundary.

B1t (7-15 cm)—Light brown (7.5YR 6/4, dry) or brown (7.5YR 5/4, moist) heavy clay loam in upper part, clay in lower part; weak medium subangular blocky structure, very and extremely hard, firm; common roots; effervesces strongly; moderately alkaline; clear smooth boundary.

B21t (15-43 cm)—Brown (7.5YR 5/4, dry) or dark brown (7.5YR 4/4, moist) clay; moderate medium and coarse blocky structure; blocks separated by an irregular network of cracks up to 0.5 cm in diameter; this material easier to remove than that below; common roots; extremely hard, very firm; effervesces strongly; clear wavy boundary.

B22t (43-75 cm)—Brown (7.5YR 5/4, dry) or dark brown (7.5YR 4/4, moist) clay; dominantly moderate coarse platy and wedgy, with some weak medium subangular blocky structure; plates and wedges nearly horizontal ranging from about 0.5-2 cm thick and up to 5 cm long; extremely hard, very firm; a few roots ranging from about 0.5-1 mm thick; slickensides on some ped faces; blocks, plates, and wedges tightly packed and difficult to remove with a hammer; a few vertical cracks ranging from < 0.5-1 mm in width extend into this horizon from the overlying horizon, and there are also cracks, commonly horizontal or nearly so, about 0.5 mm or less in width, between the structural units; effervesces strongly; moderately alkaline; clear smooth boundary.

K & B (75-90 cm)—Dominantly pink (7.5YR 8/4, dry) or light brown (7.5YR 6/4, moist) with lesser amounts of brown (7.5YR 5/4, dry) or dark brown (7.5YR 4/4, moist) clay; dominantly weak medium subangular and angular blocky, but both high-carbonate parts (light-colored) and low-carbonate parts (dark-colored) are usually arranged in roughly vertical strips ranging from 1-10 mm in diameter; a few wedges and plates, 1-2 cm thick and 2-4 cm long; very hard, firm; very few roots; a few carbonate filaments; a few hard carbonate nodules; harder than underlying horizons; effervesces strongly; moderately alkaline; clear smooth boundary.

K2 (90–112 cm)—Dominantly pinkish-white (7.5YR 8/2, dry) or light brown (7.5YR 6/4, moist) with lesser amounts of light brown (7.5YR 6/4, dry) or brown (7.5YR 5/4, moist) silty clay loam; weak medium subangular blocky structure; hard, firm; no roots; light-colored, high-carbonate zones occur in irregular masses between which are darker-colored zones lower in carbonate; few very fine tubular pores; effervesces strongly; moderately alkaline; clear wavy boundary.

Bcacs (112–133 cm)—Light brown (7.5YR 6/4, dry) or brown (7.5YR 5/4, moist) heavy clay loam; compound weak medium prismatic and weak and moderate, fine and medium subangular blocky structure; hard, firm; no roots; few carbonate nodules, 1–5 mm in diameter; powdery, fine gypsum crystals on ped surfaces in a few places; common very fine tubular pores; moderately alkaline; effervesces strongly; clear wavy boundary.

B1tcacsb (133–152 cm)—Dominantly light brown (7.5YR 6/4, dry) or brown (7.5YR 5/4, moist) with a few parts reddish-brown (5YR 5/4, dry; 5YR 4/4, moist) heavy clay loam; weak medium subangular blocky structure; hard, firm; no roots; a few powdery discontinuous gypsum coatings on ped faces; common fine and very fine tubular pores some of which are lined with reddish-brown material; effervesces strongly; moderately alkaline; clear wavy boundary.

B2tcacsb (152–163 cm)—Reddish-brown (5YR 5/4, dry; 5YR 4/4, moist) clay loam; weak fine and medium subangular blocky structure; hard, firm; no roots; black (Mn? Fe?) filamentary coatings on many ped faces; few gypsum filaments; a few carbonate filaments and nodules; many ped faces smooth and reflective; generally effervesces strongly, but a few parts noncalcareous; moderately alkaline.

Remarks—Some of the clay in the Bt horizon is thought to be of sedimentary origin, placed by movement of clay-laden water into the holes and along cracks in the Bt horizon when it is dry. However, this soil is considered to have an argillic horizon on the following evidence: (i) there is an increase in clay from A to B; (ii) although no thin sections were made of the B horizon, judging from thin sections of similar materials, the required amount of oriented clay is present; and (iii) the horizon may be traced laterally to soils of the same surface, of the same age, and only a few meters away that do have demonstrable argillic horizons and that lack the plates and wedges in the Bt horizon. Prior to development of the plates and wedges, the described soil is thought to have had an argillic horizon similar to that of the adjacent soils. Therefore, part of the clay in the present platy and wedgy horizon is considered to be of illuvial origin, placed earlier in the history of the soil.

Vertical cracks of the described width in the B21t horizon developed only after the pit had been open and the soil had dried for several days. The soil was fairly dry when the pit was dug in May 1971, with no effective rain since August 1970.

Site C

Soil Series—Stellar

Classification—Ustollic Haplargid, fine, mixed, thermic.

Soil Surface—The soil surface between grass clumps is cracked into smooth-topped polygons ranging from 1–5 cm in diameter. Cracks between polygons range from 0.5–1 mm in width. The topmost part of the polygons is readily removed and disengages as a plate ranging from about 2–8 mm in diameter. The soil surface as a whole is very smooth with virtually no microrelief.

A2 (0–5 cm)—Light gray (10YR 7/2, dry) or dark brown (10YR 4/3, moist) clay loam; weak fine and medium platy; slightly hard, very friable; few roots; few fine and very fine tubular pores; effervesces strongly; moderately alkaline; abrupt smooth boundary.

A3 (5–9 cm)—Light brown (7.5YR 6/3, dry) or dark brown (7.5YR 4/3, moist) clay loam; weak fine subangular blocky structure; hard, friable; common roots, about 1 mm in diameter; horizon becomes redder in lower part; generally noncalcareous, effervesces weakly in places; mildly alkaline; abrupt smooth boundary.

B21t (9–23 cm)—Reddish-brown (6YR 5/4, dry; 6YR 4/4, moist) clay; compound weak medium prismatic and weak medium and fine subangular blocky structure; very hard, firm; common roots, about 1 mm in diameter; sand grains coated with clay; noncalcareous; mildly alkaline; clear wavy boundary.

B22t (23–44 cm)—Reddish brown (6YR 5/4, dry; 6YR 4/4, moist) heavy clay loam; weak coarse prismatic, breaking to weak medium subangular blocky structure; very hard, firm; few roots, about 1 mm diameter; sand grains coated with clay; effervesces strongly; moderately alkaline; clear wavy boundary.

B23tca (44–67 cm)—Reddish brown (5YR 5/4, dry; 5YR 4/4, moist) clay; compound weak coarse prismatic and weak medium and coarse subangular blocky structure; very hard, firm; few roots, 0.5–1 mm in diameter; some ped faces have smooth, reflective faces; a few fine hard carbonate nodules ranging from 1–3 mm in diameter; sand grains coated with clay; effervesces strongly; moderately alkaline; clear smooth boundary.

B24tca (67–87 cm)—Reddish brown (5YR 5/4, dry; 5YR 4/4, moist) clay, with a few spots slightly redder; compound weak coarse prismatic and weak medium and coarse subangular blocky structure; very hard, firm; very few roots; sand grains coated with clay; a few fine carbonate nodules ranging from 1–5 mm in diameter; a few carbonate filaments; some ped surfaces have smooth reflective faces; effervesces strongly; moderately alkaline; clear smooth boundary.

K & Bt (87–118 cm)—About equal parts of high-carbonate material, pinkish white (7.5YR 8/2, dry) or light brown (7.5YR 7/4, moist) with parts whiter, and parts containing less carbonate, light brown (7.5YR 6/4, dry) or brown (7.5YR 5/4, moist) and reddish brown (5YR 5/4, dry; 5YR 4/4, moist) clay loam; weak medium subangular blocky structure; dominantly hard, firm; very few roots; carbonate occurs as nodular and cylindroidal forms and as masses of irregular shape with consistence ranging from slightly to very hard; effervesces strongly; moderately alkaline; clear wavy boundary.

K21 (118–134 cm)—Dominantly pinkish white (7.5YR 8/2, dry), pink (7.5YR 7/4, moist), or light brown (7.5YR 6/4, moist) with lesser amount pink (7.5YR 7/4, dry), a few stainings of pink (5YR 8/4, dry) or reddish brown (5YR 5/4, dry) heavy clay loam; weak medium and coarse platy structure; hard, firm; no roots; effervesces strongly; moderately alkaline.

Remarks—Only the upper subhorizon of the thick K horizon is described here (see pages A-91 and 92, Gile et al., 1970 for description and laboratory data for the full thickness of a similar pedon).

Site D

Soil Series—Reakor

Classification—Ustollic Calciorthid, fine-silty, mixed, thermic.

Soil Surface—The surface is cracked into polygons ranging from about 1-3 cm in diameter. Most cracks between polygons are less than 0.5 mm wide. Fragments of loose drying platelets, 1-2 mm thick, occur on the surface around the grass clumps. Tops of polygons are easily removed as plates which range from about 2-8 mm thick. A few pebbles of mixed lithology occur near the moisture blocks. Most pebbles range from about 1-2 cm in diameter; a few range up to 4 cm in diameter. There is very little microrelief in the surface as a whole.

A (0-4 cm)—Light brownish-gray (10YR 6.5/2, dry) or dark brown (10YR 4.5/3, moist) clay loam; weak fine and medium platy structure; slightly hard; few and common roots; few very fine tubular pores; effervesces strongly; moderately alkaline; abrupt smooth boundary.

A3 (4-10 cm)—Light brownish-gray (10YR 6.5/2, dry) or dark brown (10YR 4/3, moist) silty clay loam; weak coarse subangular blocky structure; very hard, firm; few fine and very fine tubular pores; common roots; effervesces strongly; moderately alkaline; clear smooth boundary.

B1 (10-26 cm)—Light brown (7.5YR 6/4, dry) or brown (7.5YR 5/4, moist) clay loam; weak fine and medium subangular blocky structure; slightly hard and hard, friable; common roots, 0.5–1 mm thick; few very fine tubular pores; effervesces strongly; moderately alkaline; clear wavy boundary.

B21ca (26-53 cm)—Light brown (7.5YR 6/4, dry) or brown (7.5YR 4.5/4, moist) silty clay loam; compound weak medium prismatic and weak fine and medium subangular blocky structure; hard, friable; few roots; few carbonate filaments; a few insect burrows, 0.5–1 cm diameter, most filled or partly filled with fine earth; effervesces strongly; moderately alkaline; clear wavy boundary.

B22ca (53-79 cm)—Light brown (7YR 5.5/4, dry; 7YR 4/4, moist) heavy clay loam; compound moderate medium prismatic and weak medium subangular blocky structure; hard, friable; few fine roots, about 0.5 mm in diameter; few fine and very fine tubular pores; effervesces strongly; moderately alkaline; clear smooth boundary.

K & B (79-95 cm)—Dominantly pink (7.5YR 7/4, dry; 7.5YR 5/4, moist) heavy clay loam, common carbonate nodules, very pale brown (10YR 9/3-7/3, dry); weak medium and fine subangular blocky structure; slightly hard and hard, friable; very few fine roots; moderately alkaline; effervesces strongly; clear wavy boundary.

B2tcab (95-122 cm)—Dominantly reddish brown (5YR 5/5 and 5YR 5/4, dry; 5YR 4/5 and 5YR 4/4, moist) with some parts light brown (7.5YR 6/4, dry) sandy clay loam; compound weak medium prismatic and weak fine and medium subangular blocky structure; hard, friable; no roots; sand grains in reddish brown parts stained with clay; common carbonate nodules and cylindroids, 0.5-2 cm diameter, ranging from slightly to very hard; reddish brown parts effervesce weakly, rest effervesce strongly; moderately alkaline; clear smooth boundary to underlying K2b horizon.

Remarks—The B2 horizon was not examined in thin section. However, judging from microscopic observations of similar horizons, it seems questionable that the 1% of oriented clay required for the argillic horizon (Soil Survey Staff, 1973) is present, and this soil is therefore considered to be a Calciorthid.

The buried soil rises to the west as the Stellar soil and is more deeply buried to the east (see Gile et al., 1970 for a discussion of the chronology and stratigraphy of soils of the basin floor in this area). The soil surface was described inside the exclosure containing the moisture blocks because of trampling by cattle outside the exclosure.

Site E

Soil Phase—Algerita, deep gypsum phase.
Classification—Typic Calciorthid, fine-loamy, mixed, thermic.
Soil Surface—Weakly crusted between grass clumps.

C (0-3 cm)—Light brown (8YR 6.5/4, dry) or brown (8YR 5/4, moist) light fine sandy loam; single grain; loose; no roots; effervesces strongly; moderately alkaline; abrupt smooth boundary.

A1 (3-12 cm)—Light gray (10YR 7/2, dry) or brown (10YR 5/3, moist) heavy fine sandy loam; massive and weak medium platy structure; hard, friable; few roots; common fine and medium tubular pores; effervesces strongly; moderately alkaline; abrupt smooth boundary.

B11 (12-31 cm)—Light gray (10YR 7/2, dry) or brown (10YR 5/3, moist)

clay loam; compound weak coarse prismatic and weak medium subangular blocky structure; very hard, firm; few roots between grass
clumps, common beneath clumps; few fine and medium tubular pores;
a few insect burrows, 0.5–1 cm in diameter, some empty and some filled
with fine earth; effervesces strongly, moderately alkaline; clear wavy
boundary.

B12 (31–50 cm)—Pinkish-gray (7.5YR 6.5/3, dry) or brown (7.5YR 5/3,
moist) light sandy clay loam; compound weak medium prismatic and
weak medium subangular blocky structure; very hard, firm; very few
roots; very few fine tubular pores; an occasional carbonate nodule,
about 0.5 cm in diameter; effervesces strongly; moderately alkaline;
clear wavy boundary.

B21ca (50–65 cm)—Pinkish-gray (7.5YR 7/3, dry) or brown (7.5YR 5/4,
moist) silty clay loam; compound weak medium prismatic and weak
fine and medium subangular blocky structure, slightly hard and hard,
friable; very few roots; few very fine tubular pores; very few carbonate
nodules ranging from 1–4 mm in diameter; effervesces strongly; moderately alkaline; clear wavy boundary.

B22ca (65–90 cm)—Pinkish-gray (7.5YR 7/3, dry) or brown (7.5YR 5/4,
moist) silty clay loam; compound moderate fine and medium prismatic
and weak fine and medium subangular blocky structure; slightly hard
and hard, friable; no roots; few carbonate nodules ranging from about
1–4 mm in diameter; effervesces strongly; moderately alkaline; clear
wavy boundary.

B23ca (90–112 cm)—Very pale brown (10YR 7/3, dry) or brown (10YR 5/3,
moist) silty clay loam; compound moderate fine and medium prismatic
and weak medium and fine subangular blocky structure; hard and slightly hard, friable; no roots; common very fine tubular pores; few carbonate nodules; effervesces strongly; moderately alkaline; clear wavy
boundary.

2C1ca (112–120 cm)—White (10YR 9/2, dry) or light gray (10YR 7/2, moist)
loam; compound weak medium prismatic and weak medium subangular
blocky structure; slightly hard and hard, friable; no roots; common
very fine tubular pores; carbonate occurs as scattered grain coatings;
some fine-grained gypsum is present; this horizon is absent in places
along the trench exposure; effervesces strongly; moderately alkaline;
abrupt smooth boundary.

2C2ca (120–138 cm)—White (10YR 8/2, dry) or pale brown (10YR 6/3,
moist) sandy loam; massive; very hard, firm; no roots; the material consists largely of fine-grained gypsum; one crack filling, nearly vertical,
consists largely of gypsum and is about 1 mm thick; common very fine
tubular pores; carbonate occurs as scattered faint patches and grain
coatings; effervesces strongly; moderately alkaline.

Remarks—The stratigraphy and chronology of the materials in and adjacent
to the playa are not known. Gypsum in the horizons from 112–138 cm
is considered to be of geologic (lacustrine) origin instead of pedologic

origin. The material from 0–112 cm is considered to be alluvium of latest Pleistocene or early to mid-Holocene age.

This soil is designated "Algerita, deep gypsum phase" to distinguish it from normal Algerita soils, which lack such gypsum deposits.

Site F

Soil Variant—Onite, buried soil variant.

Classification—Typic Haplargid, coarse-loamy, mixed, thermic.

Soil Surface—The surface is smooth and bare because of strong erosion by wind. Along the fences and around a few plants, there is an accumulation of sand, apparently of very recent origin.

B2t (0–18 cm)—Reddish-brown (5YR 5/5, dry; 5YR 4/5, moist) fine sandy loam; very weak medium subangular blocky structure; slightly hard, very friable; few roots; a few fine and very fine tubular pores; silicate clay coatings on sand grains; few termite tunnels; noncalcareous; mildly alkaline; clear wavy boundary.

B31t (18–34 cm)—Reddish-brown (5YR 5/5, dry; 5YR 4/5, moist) light fine sandy loam; massive; slightly hard, very friable; very few roots; clay coatings on sandy grains; a few termite tunnels; noncalcareous; mildly alkaline; clear wavy boundary.

B32t (34–44 cm)—Reddish-brown (5YR 5.5/4, dry; 5YR 4.5/4, moist) loamy sand; massive and single grain; soft and slightly hard, very friable and friable; very few roots; in places, tongues of sandy loam extend irregularly upward from the underlying horizon; generally noncalcareous, but effervesces weakly in a few spots; mildly alkaline; clear and abrupt, wavy and irregular boundary.

B1tcab (44–60 cm)—Reddish-brown (5YR 5.5/4, dry; 5YR 4.5/4, moist) heavy sandy loam; compound weak coarse prismatic and weak medium subangular blocky structure; hard, friable; very few roots; clay coatings on sand grains; few carbonate filaments; distinctly firmer and harder in place than the overlying horizon; commonly effervesces strongly, noncalcareous in a few places; moderately alkaline; clear wavy boundary.

B21tcab (60–76 cm)—Light reddish-brown (6YR 6/4, dry) or reddish-brown (6YR 4.5/4, moist) light sandy clay loam; compound weak coarse prismatic and weak medium and coarse subangular blocky structure; hard, friable; very few roots; a few volumes of yellowish red (5YR 5/6, dry; 5YR 4/6, moist); few dark insect burrow fillings, 0.5–1 cm in diameter; common carbonate filaments; few very fine tubular pores; clay coatings on some sand grains; most parts effervesce strongly, a few weakly; moderately alkaline; clear wavy boundary.

B22tcab (76–90 cm)—Dominantly light brown (7.5YR 6/5, dry) or brown (7.5YR 5/5, moist) light sandy clay loam; compound weak coarse prismatic and weak medium subangular blocky structure; hard and very hard, friable and firm; no roots; common carbonate nodules, white

(10YR 9/2, dry) or pink (10YR 8/4, moist); a few parts yellowish red (5YR 5/6, dry; 5YR 4/6, moist); few fine tubular pores; few carbonate filaments; sand grains in yellowish-red parts have clay coatings; few dark, fine-earth insect burrow fillings, 0.5–1 cm in diameter; most parts effervesce strongly, yellowish-red parts effervesce weakly; moderately alkaline; clear wavy boundary.

B23tcab (90–103 cm)—Dominantly reddish-brown (5YR 5.5/5, dry; 5YR 4.5/5, moist) with smaller amounts of light brown (7.5YR 6/5, dry) or brown (7.5YR 4/5, moist) light sandy clay loam; compound weak coarse prismatic and very weak medium subangular blocky structure; hard and very hard, friable and firm; no roots; a few carbonate nodules, white (7.5YR 9/2, dry) or pinkish-white (7.5YR 8/2, moist); clay coatings on sand grains in reddish-brown parts; few fine tubular pores; in places this horizon is absent and the B22tcab rests directly on the K2; most parts effervesce weakly, some parts noncalcareous; moderately alkaline; clear smooth boundary.

K2b (103–126 cm)—Pinkish-white (7.5YR 8/2, dry; 7.5YR 7/4, moist) sandy clay loam, a few parts light brown (7.5YR 6.5/4, dry) or brown (7.5YR 5.5/4, moist), and reddish-brown (5YR 5/4, dry, or 5YR 4/4, moist); weak medium and coarse subangular blocky structure; very hard, firm; few very fine tubular pores; sand grains separated by carbonate; grades to less carbonate with depth; no roots; moderately alkaline; effervesces strongly.

Remarks—Sediments from 0–44 cm are considered to be a deposit considerably younger than the materials below, and may mark a period of instability during the Holocene. The extent of this apparently younger deposit is not known. The lack of an A horizon, the Bt horizon at the surface, sand accumulations around shrubs, and sand piled along the exclosure fences—all indicate considerable wind erosion during recent years. This soil is designated "Onite, buried soil variant" because of the buried soil at shallow depths.

Site G

Soil Series—Hueco.

Classification—Petrocalcic Paleargid, coarse-loamy, mixed, thermic.

Soil Surface—A few indurated carbonate nodules, most less than 2 cm in diameter, are scattered over the surface.

C (0–5 cm)—Light brown (7.5YR 6/4, dry) or brown (7.5YR 5/4, moist) sand; loose and soft; single grain and massive; few roots; noncalcareous; mildly alkaline; abrupt smooth boundary.

A2 (5–10 cm)—Reddish-brown (6YR 5.5/4, dry; 6YR 4.5/4, moist) light fine sandy loam; massive; soft, very friable; very few roots; noncalcareous; mildly alkaline; abrupt smooth boundary.

B1t (10–23 cm)—Reddish-brown (5YR 5/4, dry; 5YR 4/4, moist) light fine sandy loam; massive; slightly hard, friable; very few roots; clay coatings on sand grains; few fine tubular pores; noncalcareous; mildly alkaline; clear wavy boundary.

B21t (23-36 cm)—Reddish-brown (5YR 5.5/4, dry; 5YR 4/4, moist) fine sandy loam; massive; slightly hard to hard, friable; harder in place than above; very few roots; clay coatings on sand grains; common fine and very fine tubular pores; noncalcareous; mildly alkaline; clear wavy boundary.

B22tca (36-46 cm)—Light reddish-brown (6YR 6/4, dry) or reddish-brown (6YR 5/4, moist) fine sandy loam; massive; slightly hard to hard, friable; very few roots; clay coatings on some sand grains; carbonate coatings on some grains, and a few carbonate filaments; few fine tubular pores; effervesces weakly and strongly; moderately alkaline; clear wavy boundary.

B3ca (46-71 cm)—Light brown (6YR 6.5/4, dry) or reddish-brown (6YR 4.5/4, moist) heavy sandy loam; very weak medium subangular blocky structure; generally slightly hard, with some parts hard, friable; no roots; common carbonate filaments and grain coatings; few fine tubular pores; effervesces strongly; moderately alkaline; abrupt wavy boundary.

K1 (71-79 cm)—Light reddish-brown (6YR 6/4, dry) reddish-brown (6YR 5/4, moist) very gravelly sandy loam; a loose mass of very fine crumbs between the indurated carbonate nodules that constitute the gravel; most nodules are less than 5 cm in diameter; nodules have pustulose surfaces; surfaces of nodules are stained reddish brown but interiors are commonly white (10YR 8/2, dry) or very pale brown (10YR 7/3, moist); effervesces strongly; moderately alkaline; abrupt wavy boundary.

K2m (79-90 cm)—Dominantly white (10YR 8/2, dry) or very pale brown (10YR 7/3, moist) carbonate-cemented material; massive; extremely hard, extremely firm; indurated; no roots; sand grains widely separated by carbonate; carbonate laminae occur discontinuously in upper part; effervesces strongly; moderately alkaline.

Remarks—Thickness of the petrocalcic horizon (the upper boundary of which is at 79 cm) is not known. However, based on exposures elsewhere, it is believed to grade through a transitional horizon with less carbonate into unconsolidated sand, possibly with some gravel, at a depth of several meters.

LITERATURE CITED

Buffington, L. C., and C. H. Herbel. 1965. Vegetational changes on a semidesert grassland range from 1858 to 1963. Ecol. Monogr. 35:139–164.

El Paso Geological Society. 1970. Cenozoic stratigraphy of the Rio Grande Valley area of Dona Ana County, New Mexico. Guidebook, 4th Annu. Field Trip, Univ. Texas at El Paso, Dep. of Geology. 49 p.

Gile, L. H. 1970. Soils of the Rio Grande Valley border in southern New Mexico. Soil Sci. Soc. Amer. Proc. 34:465–472.

Gile, L. H., and R. B. Grossman. 1968. Morphology of the argillic horizon in desert soils of southern New Mexico. Soil Sci. 106:6–15.

Gile, L. H., and J. W. Hawley. 1968. Age and comparative development of desert soils at the Gardner Spring radiocarbon site, New Mexico. Soil Sci. Soc. Amer. Proc. 32:709-716.

Gile, L. H., J. W. Hawley, and R. B. Grossman. 1970. Distribution and genesis of soils and geomorphic surfaces in a desert region of southern New Mexico. Guidebook, Soil-Geomorphology Field Conference, Soil Sci. Soc. Amer., University Park, N. M.

Gile, L. H., F. F. Peterson, and R. B. Grossman. 1965. The K horizon: A master soil horizon of carbonate accumulation. Soil Sci. 99:74-82.

Gile, L. H., F. F. Peterson, and R. B. Grossman. 1966. Morphological and genetic sequences of carbonate accumulation in desert soils. Soil Sci. 101:347-360.

Hawley, J. W., and L. H. Gile. 1966. Landscape evolution and soil genesis in the Rio Grande Region, southern New Mexico. Guidebook, 11th Annu. Field Conf., Rocky Mtn. Section Friends of the Pleistocene, University Park, N. M. 74 p.

Hawley, J. W., and F. E. Kottlowski. 1969. Quaternary geology of the south-central New Mexico border region. N. M. Bur. Mines Min. Resour. Circ. 104:52-76.

Herbel, C. H. 1963. Fertilizing tobosa on flood plains in the semidesert grassland. J. Range Manage. 16:133-138.

Herbel, C. H., F. N. Ares, and R. A. Wright. 1972. Drought effects on a semidesert range. Ecology 53:1084-1093.

Houston, W. R. 1968. Soil moisture on native grazing lands in the semiarid Northern Plains of USA. Ann. Arid Zone 7:230-234. (Published by the Arid Zone Research Ass. of India, Jodhpur.)

Kincaid, D. R., J. L. Gardner, and H. A. Schreiber. 1964. Soil and vegetation parameters affecting infiltration under semiarid conditions. p. 440-453. *In* Land erosion, precipitation, hydrometry, soil moisture. Int. Ass. Sci. Hydrol. Publ. no. 65.

King, W. E., J. W. Hawley, A. M. Taylor, and R. P. Wilson. 1971. Geology and ground-water resources of Central and Western Dona Ana County, New Mexico. N. M. Bur. Mines Min. Resour. Hydrol. Rep. 1. 64 p.

Kottlowski, F. E. 1960. Reconnaissance geologic map of Las Cruces 30-minute quadrangle. N. M. Bur. Mines Min. Resour. Geol. Map 14.

Kottlowski, F. E., R. H. Flower, M. L. Thompson, and R. W. Foster. 1956. Stratigraphic studies of the San Andres Mountains, New Mexico. N. M. Bur. Mines Min. Resour. Mem. I. 132 p.

Kramer, P. J. 1969. Plant and soil water relationships: A modern synthesis. McGraw-Hill Book Co., New York. 482 p.

Ruhe, R. V. 1967. Geomorphic surfaces and surficial deposits in southern New Mexico. N. M. Bur. Mines Min. Resour. Mem. 18. 66 p.

Russell, M. B. 1959. Interactions of water and soil. p. 35-42. *In* M. B. Russell (co-ordinator). Water and its relation to soils and crops. Academic Press, New York.

Shreve, F. 1934. Rainfall runoff and soil moisture under desert conditions. Ann. Ass. Amer. Geogr. 24:131-156.

Slatyer, R. O. 1967. Plant-water relationships. Academic Press, New York. 366 p.

Smith, B. R., and S. W. Buol. 1968. Genesis and relative weathering studies in three semiarid soils. Soil Sci. Soc. Amer. Proc. 32:261-265.

Soil Survey Staff. 1962. Supplement to Agriculture Handbook No. 18, Soil survey manual (replacing pp. 173-188). U. S. Government Printing Office, Washington, D. C.

Soil Survey Staff. 1973. Soil taxonomy, a basic system of soil classification for making and interpreting soil surveys. Soil Conservation Service, USDA. Agriculture Handbook No. 436.

Strain, W. S. 1966. Blancan mammalian fauna and Pleistocene formations, Hudspeth County, Texas. Bull. 10, Texas Memorial Museum, Austin. 55 p.

Taylor, S. A., D. D. Evans, and W. D. Kemper. 1961. Evaluating soil water. Utah Agr. Exp. Sta. Bull. 426. 67 p.

Winkworth, R. E. 1970. The water regime of an arid grassland (*Eragrostis eriopoda* Benth.) community in central Australia. Agr. Meteorol. 7:387-399.

Oxygen Content in the Ground Water of Some North Carolina Aquults and Udults[1]

8

R. B. DANIELS, E. E. GAMBLE, AND S. W. BUOL [2]

ABSTRACT

Measurement of oxygen content in Aquults and Udults indicates that the Aquults have reducing conditions (Fe^{2+} is stable) within 1 m of the surface that may last for months. The associated Udults have reducing conditions of short duration generally at depths of 2 m or more below the surface. However, many Udults are water saturated within 20 cm of the surface, but oxygen levels of the ground water remain high enough so that iron reduction does not take place. Aquults can be saturated to or above the soil surface for 4 to 5 months, yet the oxygen levels in the upper 0.5 m may remain high enough so that oxidizing conditions (Fe^{3+} is stable) exist. Thus water saturation does not always mean that reducing conditions prevail, even though organic rich A horizons are saturated.

INTRODUCTION

The Coharie geomorphic surface near Newton Grove, North Carolina, is a gently undulating tableland with a local relief somewhat less than 1 m. Udults, soils with brown B horizons, occur on the slightly convex parts of the tableland and other areas where water does not stand above the surface. Aquults, soils with gray B horizons, occur in the concave areas and on broad flats where water does not run off readily. The water table is deep near the dissected edge of the tableland but it is close to the surface near the centers of the flats. Aquults frequently occur near the edge of the tablelands where summer and fall water tables may be below 3 m (10 feet). The gray colors in these soils have been interpreted as indicating wet conditions regardless of where they occur on the landscape. Field workers seldom challenge this interpretation because the Aquults do stay wet longer than the Udults. But Aquults near the dissected edge of the tableland have water table regimes similar to the adjacent Udult (Table 1). In fact, the mean water table may

[1] Paper no. 3488 of the Journal Series. Joint contribution from the Soil Conservation Service, U. S. Department of Agriculture, and the Department of Soil Science, North Carolina Agricultural Experiment Station, Raleigh, North Carolina.

[2] Soil Scientists, Soil Conservation Service, and the Soil Science Department, North Carolina State University, Raleigh, North Carolina.

[3] Professor, Department of Soil Science, North Carolina State University, Raleigh, North Carolina.

Table 1. Mean water table depths of Aquults near Newton Grove, North Carolina.

Time of year	Distance of site from dissected edge of geomorphic surface			
	0.18 km Aquult	0.18 km Udult	0.24 km Aquult	0.45 km Udult
	water table depths, cm			
1st quarter	72 (-10)*	89	24	44
2nd quarter	97 (+5)	129	85	81
3rd quarter	247 (+17)	291	104	110
4th quarter	251 (+14)	292	132	117
Yearly	167 (+7)	201	87	85

* Difference, in centimeters, between the water table in the Aquult and the Udult at 0.18 km.

be deeper in the Aquult than it is in some Udults located farther from the dissected edge.

Soil scientists have generally associated the gray colors, (chroma, two or less) in soils such as the Aquults with reducing conditions. The iron is believed to be reduced when the horizons are water saturated and organic matter is present to provide an energy source for microorganisms. If reducing conditions occur in some water-saturated horizons, but not in others, differences between these horizons should be measurable directly by taking oxidation-reduction (Eh) measurements or indirectly by measuring the oxygen content of the ground water. Portable oxygen meters make it feasible to monitor the oxygen content of the ground water in wells used for water table studies. The present study reports on the relations between oxygen content of ground water and horizon color.

METHODS

An oxygen meter (Yellow Springs Instrument Company Model no. 51) was used to measure the oxygen content of the soil ground water. The instrument was calibrated in air saturated with water vapor. This was done in a calibration chamber with a small amount of water in equilibrium with the air in the chamber. The oxygen probe was inserted in the chamber and the system allowed to cool to approximately test measurement temperature. The instrument was set for the amount of oxygen contained in the air at the calibration temperature. An altitude correction and an adjustment between calibration and sample temperatures were made.

The oxygen content of the soil ground water was measured in wells cased with a 2.54-cm diameter thin wall (EMT) electrical conduit. Oxygen measurements were made in two different sets of wells installed at three sites representing the wet, intermediate, and dry soil environments near Newton Grove, North Carolina. These two sets of wells were used to follow in detail the oxygen content of the ground water. The casings were perforated in the lower 15 cm and set at depths of equal intervals, or in major soil horizons. In addition, wells used to follow water table levels for another study were sampled.

Wells were drained and allowed to refill with at least 20 cm of water before temperature and oxygen measurements were made. Draining was necessary because oxygen measurements in the undrained wells were different from those of wells that were pumped dry and allowed to refill. In 41 checks, 21 wells had increases and 20 wells had decreases in oxygen in the water entering after pumping. The fresh water that entered the wells after pumping is believed to more closely represent the oxygen status of the soil water at the depth of the well perforations. The 2.54-cm well diameter required only about 1,500 cm³ of water to fill a casing 2 m long, and the oxygen measurement was always taken at the bottom of the well. Therefore, only the first water entering the well would be exposed to the air, and the surface area in contact with the air is small in comparison to the water volume. Thus, there probably is no increase in oxygen in the water at the base of the well because oxygen diffusion rates in water are low (Howeler & Bouldin, 1971).

RESULTS

The oxygen content of ground water is of interest only in so far as it reflects the oxidation-reduction regime of the soil because most soil morphologists believe that it is this regime that controls soil color and mottling patterns. We define reducing conditions as the Eh-pH combinations that are in the stability field of ferrous iron (Fig. 1) and oxidizing conditions as those that are in the stability field of ferric hydroxide. Note that Fe may be reduced at positive Eh values and that the reaction is pH dependent.

Oxygen concentrations can be converted to Eh values by the use of partial pressures. An oxygen concentration of 0.1 ppm represents a partial pressure of about 10^{-6}. At pH 5 the ferric hydroxide-ferrous iron stability boundary (Fig. 1) requires an Eh value of about +0.3 volts and an oxygen partial pressure of about 10^{-40} (Garrels & Christ, 1965), or an oxygen partial pressure about 31 orders of magnitude less than found at 0.1 ppm.

The accuracy of the oxygen meter we used is ± 0.25 ppm and readability is 0.1 ppm. Therefore, the meter will not read the low oxygen values that would be encountered in actual reducing conditions. But when oxygen measurements consistently show values of 0.2 ppm or less, it is probable that oxygen levels may be low enough to have a stable ferrous iron environment. To test this assumption, the following laboratory tests were made to correlate the oxygen level read in the field with possible reducing conditions.

About 250 g of Cecil B horizon, a red (2.5YR 5/6) clayey Ultisol, were placed in a 1-liter Erlenmeyer flask with 85 g of sugar in 1 liter of water. The sugar solution was used in runs 1 and 2 and water only in run 3. The flask was stoppered to prevent oxygen in the air from entering the solutions. Daily oxygen measurements were made with the oxygen meter calibrated as it was for field work. Eh and pH measurements were made at the same time after the first 5 days in run 1 and for each day in runs 2 and 3. The results are given in Table 2 and are graphically shown in Fig. 1.

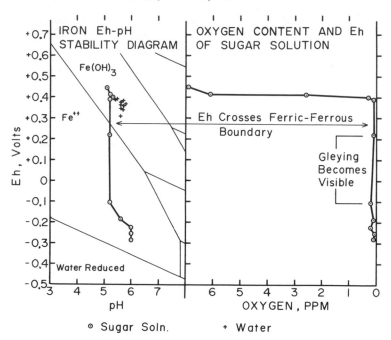

Figure 1. Iron Eh-pH stability diagram and oxygen content and Eh of sugar solution above a Cecil B horizon.

During the first few days there was little change in Eh, but there was a large loss of oxygen even in the run using only water (Fig. 1). Once the oxygen values were reduced to 0.2 ppm or less, the Eh decreased rapidly. From data in Table 2 and Fig. 1, we conclude that oxygen contents of 0.2 ppm or less, as recorded by the oxygen probe, represent reducing conditions or conditions that are at least very close to the ferric hydroxide-ferrous iron stability boundary within the pH range of the soils studied (Collins & Buol, 1970). Note also from the comments in Table 2 that this point corresponds to the development of gley colors in the soil verifying Fe^{3+} to Fe^{2+} transformations.

Initial oxygen measurements were made in wells established for the purpose of characterizing the yearly water table regime of many kinds of soils. The first results indicated that mean oxygen content of ground water in Aquults was lower than it was in the Udults at all depths (Table 3). Medium-textured soils (sandy clay loam to sandy clay) have lower means than sandy soils (loamy sand to sand) of comparable drainage.

We then installed several wells at three sites about 20 m apart near Newton Grove, North Carolina, to follow in detail the seasonal oxygen changes. The data obtained for 5 consecutive months, representing 13 measurements, for two of the soils are shown in Fig. 2. Data for the Aeric Paleaquult are not shown in Fig. 2, but they are intermediate between the two soils shown.

The high oxygen content of the Typic Paleaquult during the first part of the measurement period in January 1967 probably is the result of a dry

Table 2. Changes in oxygen content, Eh, and pH of water above a Cecil B horizon with time after additions of sugar solutions of distilled water.

Day of trial	Temperature, °C	Oxygen, ppm	Eh	pH	Comments
			Run No. 1 (85 g of sugar in 1 liter of water)		
0	25.0	8.0	--	--	
1	20.7	5.7	--	--	
2 (am)	18.8	3.8	--	--	
2 (pm)	17.7	3.6	--	5.3	
3	17.0	2.4	--	--	
4	24.2	0.3	+0.401	5.8	
5	25.8	0.2	+0.451	5.2	Added 35.5 g sugar
6	26.0	0.1	+0.196*	5.2	Gley spot (N4/0)
7	22.2	0.2	+0.256*	5.1	2 gley areas size of a dime
8	25.6	0.1	+0.186*	5.2	
9	25.7	0.1	-0.074*	5.6	75-80% of soil gleyed
10	26.0	0.1	-0.189*	5.8	100% gleyed, water above soil is turbid
			Run No. 2 (85 g of sugar in 1 liter of water)		
0	--	6.9	+0.446	5.1	
1	--	6.1	+0.416	5.2	
2	--	2.6	+0.416	5.3	
3	--	0.3	+0.401	5.3	
4	--	0.1	+0.391	5.2	
5	--	0.1	+0.221	5.2	No real sign of gleying
6	--	0.2	-0.104	5.2	15-20% of soil is gleyed
7	--	0.1	-0.184	5.6	90% gleyed
8	--	0.2	-0.224	6.0	100% gleyed and turbid solution
9	--	<0.1	-0.254	6.0	
10	--	0.1	-0.284	6.0	
			Run No. 3 (no sugar added)		
0	--	7.2	+0.391	5.5	
1	--	5.7	+0.386	5.4	
2	--	2.4	+0.386	5.7	
3	--	1.5	+0.366	5.8	
4	--	1.0	+0.376	5.6	
5	--	0.7	+0.361	5.6	
6	--	0.7	+0.361	5.6	
7	--	0.8	+0.341	5.6	No visible sign of gleying after 30 days
8	--	1.4	+0.346	5.7	
9	--	1.6	I0.351	5.6	
10	--	1.6	+0.361	5.7	

* Reducing conditions, or into the Fe^{2+} stability area at the pH indicated (from Collins & Buol, 1970).

fall in 1966. The water levels of all soils in the area were below 3 m in late summer and fall but rose to a maximum height during a 2-week period in early January 1967. The Udult maintained a high oxygen concentration in the ground water throughout the measurement period. The oxygen levels in the Aquult below 1 m decreased to 2 ppm or less shortly after initial wetting, and there were short periods when oxygen contents were less than 0.2 ppm. Although the Aquult was colder than the Udult, water temperatures throughout the period were high enough to support biological activity. This set of wells was destroyed by a bulldozer after 5 months of data collection.

Another set of three groups of wells, set 2, was installed in late 1967 in an area similar to that of the first set. The wells were placed to depths of 0.6, 1.2, 1.8, 2.4, 3.0, and 4.5 m. Figures 3, 4, and 5 show the variation in

Table 3. Mean oxygen contents of ground water as related to soil drainage and texture.

Depth, m	Mean oxygen concentration, ppm	
	Medium-textured soils	
	Udults	Aquults
0 - 1.8	2.87 ± 2.6 (n=70)	1.06 ± 1.7 (n=210)
1.8 - 3.6	2.58 ± 1.3 (n=107)	0.97 ± 1.3 (n=173)
3.6	3.58 ± 2.8 (n=37)	0.40 ± 0.5 (n=48)
	Sandy soils	
	Spodic quartzipsamment	Haplaquod
0 - 1.8	4.24 ± 2.5 (n=29)	2.52 ± 2.2 (n=50)

Analysis of variance

Drainage $F_{1,720} = 150.9**$
Texture $F_{1,720} = 69.3**$
Interaction $F_{1,720} = 215.7**$

** Significant at 0.01.

oxygen content and water temperature of an Aquic Paleudult, and an Aeric and Typic Paleaquult located about 10 m apart. A total of 22 measurements in each cased hole was made over a 20-month period.

The water in the A or upper B horizons has more oxygen than it does in the deeper horizons. But among the soils at any one time period, the oxygen content generally is higher in the Udults than it is in the adjacent Aquults. The decrease in oxygen content from greater than 0.2 ppm in the Aquults in November 1969 occurs almost immediately after the first major rainfall following leaf fall. This abrupt decrease in oxygen concentration suggests a large increase in biological activity was produced by large quantities of water-soluble carbon compounds that were removed from the freshly fallen leaves. At a given depth, the Aquults have less than 0.2 ppm oxygen for much longer periods than the adjacent Aquic Paleudult (Fig. 3, 4, 5, and Table 4).

The isotherms of Fig. 3, 4, and 5 have the same general pattern of either warming or cooling from the surface downward. The isotherms are displaced to the right, or later in time, with greater depth. The wettest soil, the Typic Paleaquult, has lower temperatures than the driest soil, the Aquic Paleudult (Table 5). The slope of the isotherms to the right with depth, or the delay in warming or cooling, probably is related to a net downward water movement as well as to thermal conductivity.

INTERPRETATIONS

Interpretation of the oxygen data must be made in reference to the control of the water table that is exerted by the local geology. The surficial sediments in the Newton Grove area are sandy clay loam near the surface and grade downward to sandy loam or loamy sand at the base. These loamy sedi-

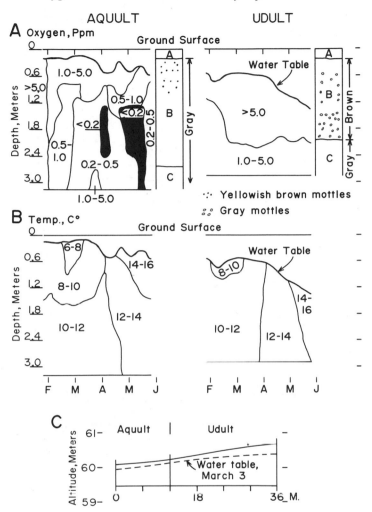

Figure 2. Oxygen content (*A*), temperature (*B*) of ground water, and landscape relations of Aquults and Udults, Set 1.

ments overlie less permeable marine clays and silty clay loams. The contact between the surficial and underlying sediments is nearly level with a gentle southeast slope of 18 to 36 cm/km (1 to 2 feet/mile). Slopes near the stream valleys truncate the contact between the marine clays and surficial sediments. The surficial sediments are only 6 to 7 m thick, and it is 1 km or more between adjacent streams. Thus there is a relatively thin permeable layer overlying nearly impermeable sediments with a water table perched above the contact.

Water movement through these surficial sediments should be similar to that of a drain tile resting on an impermeable bed (Edwards, 1956); in our

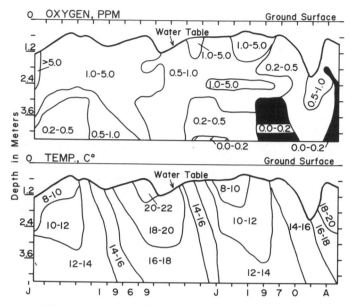

Figure 3. Oxygen content and temperature of ground water in an Aquic Paleudult, Set 2.

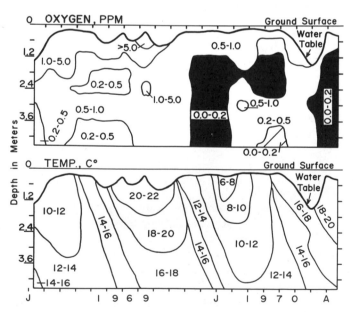

Figure 4. Oxygen content and temperature of ground water in an Aeric Paleaquult, Set 2.

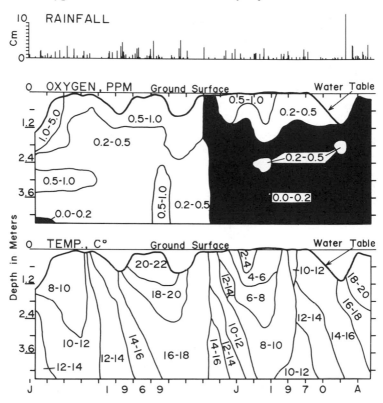

Figure 5. Rainfall, oxygen content, and temperature of a Typic Pale-aquult, Set 2.

case, the impermeable layer would be the marine clays. The flow streamlines in the surficial sediments should be down through the fine-textured upper material. The major horizontal flow should be in the coarse-textured basal sediments overlying the marine clays. But this horizontal flow will be only a few centimeters/day because the water table drops only about 1 cm/day from aquifer leakage. It is doubtful that there is much horizontal movement in the upper part of the water table because the hydraulic head differences between adjacent dry and wet soils are small (Table 1).

All water tables in the area drop during periods of no rainfall in the winter, even when vegetation is dormant and water loss by transpiration is at a minimum. This net downward movement of the water table, or aquifer leakage, is about 1 cm/day for all soils when the water table is about 4.5 m. Aquifer leakage probably is of some importance in determining soil morphology because once a material is brought into solution or suspension it is possible to remove it from the soil system via leakage. Precipitation reactions must take place while the ground water is moving through the solum, or most reactants will be lost from the soil system.

Table 4. Percentage of time the oxygen concentration of ground water is within given limits.

Drainage	Depth, m	Oxygen content in ppm					
		0.0-0.2	0.2-0.5	0.5-1.0	1.0-5.0	5.0	Dry
		%					
Set 1 (5 months data, 1967)							
Aquic	0.6	--	--	--	--	41.2	58.8
Paleudult	1.2	--	--	--	7.5	83.1	9.4
(moderately well)	1.8	--	--	--	17.7	82.3	--
	2.4	--	--	--	--	100.0	--
Aeric	0.6	--	--	--	90.0	1.8	8.2
Paleaquult	1.2	--	--	6.9	91.3	1.8	--
(somewhat poorly)	1.8	--	--	--	98.2	1.8	--
Typic	0.6	--	--	15.2	81.7	3.1	--
Paleaquult	1.2	--	32.9	50.6	13.4	3.1	
(poorly)	1.8	25.9	46.8	18.6	6.9	1.8	--
	2.4	17.7	56.3	22.2	--	3.8	--
	3.0	--	67.2	15.8	12.6	4.4	--
Set 2 (20 months data, 1969-1970)							
Aquic	0.6	--	3.4	8.0	23.9	--	64.7
Paleudult	1.2	--	10.0	26.2	52.3	--	11.5
(moderately well)	1.8	1.8	16.5	39.7	37.1	1.5	3.4
	2.4	1.3	5.4	27.8	64.4	1.1	--
	3.0	10.0	19.8	55.7	14.5	--	--
	4.5	30.7	37.4	37.4	28.9	3.0	--
Aeric	0.6	--	13.4	31.7	25.5	5.7	23.7
Paleaquult	1.2	25.5	2.8	39.4	27.1	--	5.2
(somewhat poorly)	1.8	37.1	25.0	28.9	9.0	--	--
	2.4	29.1	28.4	29.6	2.9	--	--
	3.0	21.9	23.7	54.4	--	--	--
	4.5	32.2	42.5	25.3	--	--	--
Typic	0.6	10.6	23.2	42.8	4.4	--	19.0
Paleaquult	1.2	42.0	36.1	15.2	5.7	--	1.0
(poorly)	1.8	45.5	36.6	15.5	2.5	--	--
	2.4	43.0	57.0	--	--	--	--
	3.0	48.7	36.8	22.7	--	--	--
	4.5	54.4	40.5	5.1	--	--	--

When saturated with water, the Aquults may have oxygen concentrations low enough for reducing conditions to exist within the sola for several months (Fig. 4, 5). Ferrous iron would be stable within these reduced horizons, but probably would not remain because the water in the saturated zone moved down and out of the solum. The soils studied are all pre-Pleistocene (Daniels et al., 1966), and it is probable that the water regimes have been similar to the present one for the last 1 or 2 million years. The movement of water by aquifer leakage has provided ample opportunity for removal of large quantities of ferrous iron from horizons subjected to periodic reduction. The horizons of Aquults that are subject to periodic reducing conditions are gray (10YR 6/2), as are a number of the deeper horizons of some Udults. These gray horizons generally are iron poor and contain low quantities of dithionite extractable iron (Daniels, Simonson, & Handy, 1961; Table 6).

Table 5. Percentage of time that water temperature is within given limits. Data from Set 2, 1969-1970.

Drainage	Depth, m	Temperature in degrees centigrade										
		2-4	4-6	6-8	8-10	10-12	12-14	14-16	16-18	18-20	20-22	Dry
Aquic	0.6	--	--	--	17.3	9.0	3.9	--	--	--	5.1	64.7
Paleudult (moder-	1.2	--	--	--	20.4	12.6	9.5	16.2	4.6	22.5	5.7	8.5
ately well)	1.8	--	--	--	2.6	30.9	12.6	17.0	7.2	26.3	--	3.4
	2.4	--	--	--	1.3	29.9	14.7	17.0	15.7	21.4	--	--
	3.0	--	--	--	1.3	16.2	29.6	17.3	36.9	--	--	--
	4.5	--	--	--	--	--	55.9	22.7	21.4	--	--	--
Aeric	0.6	--	--	5.4	7.2	20.1	8.2	6.2	7.7	5.9	15.6	23.7
Paleaquult (some-	1.2	--	--	1.0	10.0	24.2	9.0	10.6	9.5	15.6	14.9	5.2
what poorly)		--	--	--	8.5	26.8	10.6	13.9	9.3	30.9	--	--
	2.4	--	--	--	--	32.5	15.7	17.5	13.1	21.2	--	--
	3.0	--	--	--	--	24.7	25.0	17.0	22.2	11.1	--	--
	4.5	--	--	--	--	--	51.0	24.7	24.3	--	--	--
Typic	0.6	2.8	9.5	1.8	14.2	5.4	3.6	14.4	5.9	6.2	17.2	19.0
Paleaquult (poorly)	1.2	--	4.1	8.0	18.8	1.8	9.5	15.2	13.1	28.5	--	1.0
	1.8	--	--	11.1	18.8	10.6	14.4	10.3	19.3	15.5	--	--
	2.4	--	--	4.4	19.6	14.7	19.6	12.1	25.2	4.4	--	--
	3.0	--	--	15.7	25.5	17.3	17.0	24.5	--	--	--	--
	4.5	--	--	4.6	29.9	26.0	22.4	17.1	--	--	--	--

The horizon color is largely the color of bare quartz grains and iron-poor clay. Although reducing conditions may occur in these horizons, the quantity of ferrous iron in solution probably is low because there is little iron that can be brought into solution.

The rainfall, temperature, and oxygen levels of the Typic Paleaquult shown in Fig. 5 have several interesting interactions. During most of 1969, the oxygen levels did not become low enough to suggest that reducing conditions existed. This occurred even when the water table was within the A1 horizon and the temperature was 20 to 22C. Leaf fall in the study area occurs from about mid-October to the first part of November. Approximately 5 cm of rain about the first of November was followed by reducing conditions throughout the water-saturated zone although water temperatures were 12C or less, so biological activity would not be at its maximum. Following subsequent rains, the oxygen levels in the upper meter increased and remained above 0.2 ppm until August when a rainfall of about 15 cm was followed by reducing conditions throughout the saturated zone.

We believe that the sharp drop in oxygen levels is produced in both cases cited above by biological oxidation of large quantities of water-soluble carbon compounds entering the soil system. In November 1969, the initial rain following leaf fall was large, and the water tables were near 1 m. The water-soluble carbon initially leached from the leaves was put into a saturated soil rather than into one that was brought only to field capacity. Under these conditions, microbial activity would be under saturated conditions where oxygen diffusion rates are only about one-ten thousandth the diffusion rate through air-filled pores (Howeler & Bouldin, 1971). The heavy rainfall in August 1970, would result in some runoff from the slightly higher Udults to

Table 6. Organic carbon, extractable iron, and pH of Aquults and Udults near Newton Grove and Benson, North Carolina.

Depth	Matrix color	Organic carbon	Extractable iron	pH (1:1) H$_2$O
cm		%	%	
		Aquult (S65NC82-3)*		
0-8	10YR 3/1	2.00	0.06	3.9
8-13	10YR 5/2	0.22	0.08	4.3
13-31	10YR 6/1	0.23	0.43	4.3
31-58	10YR 6/1-6/6m[†]	0.19	0.83	4.4
58-84	10YR 6/1-2.5YR 4/8m	0.19	1.50	4.4
Tongues	red		3.60-2.20	
84-194	yellow		3.60-0.56	
	grey		0.25-0.07	
194-214C	10YR 5/1	0.05	0.15	4.2
214-244C	10YR 5/1	--	0.36	4.3
		Aquult (65NC50-2)		
0-10	5YR 2/1	3.91	0.15	4.0
10-23	10YR 6/1	0.51	0.11	4.2
23-41	10YR 7/1 5/6m	0.15	0.21	4.3
41-74	10YR 6/1 5/6m	0.13	1.20	4.4
74-124	10YR 6/1 5/8m	0.13	1.60	4.4
124-160	10YR 6/1 5/8m	0.11	0.53	4.3
160-224	10YR 6/1	0.06	0.90	4.7
224-265C	2.5 7/2	0.02	0.02	4.5
		Udult (S65NC82-1) unit 10		
0-25	10YR 3/2	0.78	0.32	4.8
25-53	10YR 5/4	0.19	0.97	4.8
53-86	10YR 5/4	0.07	1.50	4.9
86-117	10YR 6/6-7/2m	0.04	2.00	4.9
117-172	10YR 5/6-2.5 5/2m	0.05	1.60	5.0
172-244	10YR 4/6-7m	0.02	1.90	4.6
244-287C	10YR 5/8 7/1	0.02	0.57	4.5
		Udult (S65NC51-3)		
0-13	10YR 4/2	0.89	0.42	4.9
13-23	10YR 6/3	0.38	0.33	4.9
23-51	10YR 5/6	0.24	2.60	4.7
51-68	10YR 5/6	0.19	2.80	4.6
68-84	10YR 5/8	0.10	2.50	4.7
84-109	10YR 5/6	0.08	2.80	5.1
109-137pl[‡]	10YR 5/6-2.5YR 4/8m	0.06	3.30	4.9
137-162	10YR 6/6	0.04	2.10	4.9
162-228	10YR 5/6	0.06	1.40	4.9
228-274C	10YR 6/6 7/2	--	0.59	4.7

* Soil Survey Investigation Laboratory, Beltsville, Maryland, profile number.
† m = mottles.   ‡ pl = plinthite.

the Aquult. It is probable that large quantities of water-soluble carbon entered the Aquult, especially since June and July were dry months.

We do not want to infer that water-soluble carbon does not enter the soil at other times. Unpublished data from sandy soils show that each rain penetrating to 10 cm apparently carries some water-soluble carbon into the soil, and about four times more carbon through the A horizon of the wet soils than the A horizon of the drier soils. Part of this difference in carbon can be related to a greater accumulation of organic matter in the A horizon (Table 6), longer periods of saturation of the A horizon, and more total production of carbon at the wet than the dry sites.

Rainwater is essentially oxygen saturated when it reaches the ground, 10 or more ppm, but in all soils almost 90 to 95% of the oxygen is lost when the rainfall penetrates to the ground water (Fig. 2 to 5). Apparently, water-soluble carbon is not present in sufficient amounts, or in readily usable forms, to support biological activity large enough to consume all the oxygen. Apparently, this is true even when the organic-rich A horizon is water saturated for long periods. Almost all of the carbon entering these medium-textured soils must be eventually oxidized because they have no subsurface horizons of carbon accumulation similar to the sandy soils with Bh horizons.

The relatively high oxygen content of the upper layers of ground water in the Udults when compared to Aquults (Table 4) may also be related to factors other than total water-soluble carbon entering the soil system. In most Udults the water table is usually 50 cm or more deep, whereas in the Aquults it is frequently above this depth (Fig. 3, 4, and 5). Most rainwater entering the Udults must penetrate at least the upper part of the B horizon before reaching the ground water. Water collected 10 cm below the surface of sandy soils has the characteristic tea color of the "black" rivers in the Southeast. Passing this water through an ordinary paper filter will remove much of the color. It would seem that the B horizon could have much the same effect on water-soluble carbons as filter paper. If true, then much of the carbon would be trapped in the upper B horizon where it could be oxidized under aerobic conditions. The carbon remaining in the water would be oxidized under saturated conditions and would lower, but not necessarily deplete, the oxygen of the ground water. The same would apply to Aquults when water tables are deep, but not when water tables are above the B horizon. Thus, on the average, much of the water-soluble organic carbon in Aquults may be oxidized below the water table where anaerobic conditions have a chance to develop.

SUMMARY

The oxygen content of ground water in and under Aquults usually becomes low enough sometime during the year to produce reducing conditions. Such reducing conditions seldom occur in the upper 2 m of the Udults studied although some horizons of these soils are saturated for long periods. Water-soluble organic carbon probably is the energy source for microorganisms involved in oxygen consumption in saturated horizons.

Soil scientists have used gray colors in the B horizon and thick dark A horizons as indicators of wetness in soils. It has been postulated that reducing conditions are responsible for the gray colors in poorly drained soils and that oxidizing conditions are responsible for the brighter colors in the better-drained soils. There is considerable basis for this belief (Ignatieff, 1941;

Quispel, 1947; McKeague, 1965; Bonner & Ralston, 1968), and our work has reinforced these ideas. About the only modification we want to make in this general theory is to emphasize the fact that saturation alone does not lead to reducing conditions; there must also be an abundant energy source available. We also want to emphasize that gray colors (10YR 7/2 to 6/2) probably represent iron-poor conditions, but in Aquults this color probably has been formed by reduction and removal of iron not by reduction alone.

Under North Carolina conditions, we doubt that soils can be properly classified into Udults and Aquults based only on depth and duration of the water table, although these criteria apparently have been successfully applied in Pennsylvania (Latshaw & Thompson, 1968). The depth of gley mottles may be a guide in classification, but it should be used with caution because these gley mottles may represent past conditions, not those currently influencing the soil. Conversely, a soil can be water saturated and still have bright colors if the oxygen content of the water remains high enough so that the Eh remains in the ferric hydroxide stability field. However, there is a strong relation between classification units and oxidation-reduction regimes of the soils we studied. Udults, even when water saturated in the B horizon, have only short periods of reduction at depths of 1 to 2 m. Aquults can have long periods when reducing conditions exist at this depth and yearly periods of 1 to 2 months when reducing conditions occur above 1 m.

LITERATURE CITED

Bonner, F. T., and C. W. Ralston. 1968. Oxidation-reduction potential of saturated forest soils. Soil Sci. Soc. Amer. Proc. 32:111–112.

Collins, J. F., and S. W. Buol. 1970. Effects of fluctuations in the Eh-pH environment of iron and/or manganese equilibria. Soil Sci. 110:111–118.

Daniels, R. B., E. E. Gamble, W. H. Wheeler, and W. D. Nettleton. 1966. Coastal plain stratigraphy and geomorphology near Benson, North Carolina. Southeast. Geol. 7: 159–182.

Daniels, R. B., G. H. Simonson, and R. L. Handy. 1961. Ferrous iron content and color of sediments. Soil Sci. 91:378–382.

Edwards, D. H. 1956. Water tables, equipotentials, and streamlines in drained soils with anistropic permeability. Soil Sci. 81:3–18.

Garrels, R. M., and C. L. Crist. 1965. Solutions, minerals, and equilibria. Harper and Row, New York.

Howeler, R. H., and Bouldin, D. R. 1971. The diffusion and consumption of oxygen in submerged soils. Soil Sci. Soc. Amer. Proc. 35:202–208.

Ignatieff, V. 1941. Determination and behavior of ferrous iron in soils. Soil Sci. 51: 249–263.

Latshaw, G. J., and R. F. Thompson. 1968. Water table study verifies soil interpretations. J. Soil Water Conserv. 23:65–67.

McKeague, J. A. 1965. Relationship of water table and Eh to properties of three clay soils in the Ottawa Valley. Can. J. Soil Sci. 45:49–62.

Quispel, A. 1947. Measurement of the oxidation reduction potentials of normal and inundated soils. Soil Sci. 63:265–275.

Hydrology and Soil Science[1]   9

C. R. AMERMAN[2]

ABSTRACT

With advances in mathematical and computer sciences, hydrologists are reaching out from the traditional empirical-statistical approach. Among other goals, they are trying to delineate physical processes and to describe them using mathematical physics. Soil constitutes an important part of the physical framework of most hydrologic systems. Physics-based approaches require soil water characteristic curves and relationships between hydraulic conductivity and water content. Other needs include mapping soils in terms of their hydraulic characteristics. Less rigorous approaches require various other data such as infiltration curves and water storage availability above wilting point. Whatever the approach, hydraulic properties of soils will certainly be needed as input data. Besides providing soil data, soil scientists are in a position to aid hydrologists in understanding many of the water flow mechanisms of the soil. An understanding of terrestrial hydrology may in turn contribute to an understanding of some problems of soil science.

INTRODUCTION

Water and soil have been managed and studied since ancient times, and around each has developed a specialized scientific discipline. Each discipline incorporates some of the subject matter of the other, although the need for soil study has not been generally as well recognized among hydrologists as the need for water study among soil scientists.

Water movement in the unsaturated zone of the earth's mantle has been practically ignored in hydrology, which has two main branches—surface water and ground water. The former term applies to water flowing over the soil surface and in streams, and the latter refers to water below the water table. Consideration of water movement in the hydrologically neglected aerated soil and rock zone between land surface and water table may help to significantly improve runoff prediction and control, especially for small watersheds. Also of significance today, hydrologic investigations of the unsaturated zone should provide knowledge of subsurface water flow paths and velocities. This is a prerequisite to the movement of dissolved pollutants and the progress of the various chemical and biological changes to which they are subject as they pass through soil and rock.

[1] Contribution from the North Central Region, Agricultural Research Service, U. S. Department of Agriculture, Madison, Wisconsin, in cooperation with the Wisconsin Agricultural Experiment Station.

[2] Research Hydraulic Engineer, Agricultural Research Service, U. S. Department of Agriculture and Assistant Professor of Soils and Agricultural Engineering, University of Wisconsin, Madison.

It is the purpose of this paper to introduce soil scientists to the subject of hydrology. We shall discuss the author's view of the present status of the science and will discuss in detail a newly emerging branch, physics-based hydrology. Soil physicists have already contributed much to the development of this branch. It seems likely that soil surveyors, morphologists, soil chemists, microbiologists, plant physiologists, and other soil science and agronomic specialists also have knowledge and insights to contribute.

STATUS OF HYDROLOGY

Some perspective of hydrology, its background, and present status may help to show where soil science inputs are needed. Hydrology is the study and management of water as it occurs and moves about the earth. Some modern definitions include the chemistry of water within the purview of hydrology. In the strictest sense, all waters of the earth—atmospheric, oceanic, and terrestrial are in the domain of hydrology. However, the hydrology discussed here is limited to that associated with the terrestrial earth.

Modern terrestrial hydrology, though having ancient antecedents, had its beginnings in the 17th century (Biswas, 1965). However, development was slow until an upsurge began in the 1920's.

The first approaches were mainly *empirical* and consisted of studies of the various individual mechanisms thought to be influencing runoff (Lowdermilk, 1926; Sherman, 1932; Horton, 1935; Barnes, 1939). Examples of these mechanisms are: interception by vegetation, infiltration, storage of water in the aerated zone of the soil, and routing of overland flow and streamflow. These investigations, which are still being pursued, have identified a number of physical mechanisms and have produced one or more mathematical formulations for each of them.

As *statistical* theory developed and became available to hydrologists, attempts were made to document cause and effect relations and to develop regression equations for the investigation of influences upon and the prediction of runoff (Ford, 1959; Harrold et al., 1962; Snyder, 1962; Wicht, 1965; Schreiber & Kincaid, 1967; Mustonen, 1967; Veitch & Sheperd, 1971). It soon became apparent (i) that, since a description of precipitation must involve principles of probability, many facets of hydrology also involve probability components, and (ii) that statistical methods can be a valuable tool for such tasks as finding the range of magnitudes and frequency of occurrence of storms, floods, and low flows.

An offshoot of the statistical approach, *stochastic* (also called *synthetic*) hydrology, receives the attention of an important segment of the hydrologic community today (Chow & Rasmaseshan, 1965; Committee on Surface Water Hydrology, 1965; Beard, 1967; Grace & Eagleson, 1967; Matalas, 1967; Wiser, 1967; Fiering & Jackson, 1971). In these studies, statistical parameters of precipitation and streamflow are evaluated for artificially generating long periods of precipitation or streamflow data having parameters numerically the same as those for the natural data. The artificial data predict in a general way the properties of precipitation or streamflow that can be expected in the future, provided changes in the hydrologic system are minor, and are used for many purposes in water resources management.

While statistical and empirical studies often try to explain hydrologic mechanisms, stochastic models of hydrology take no cognizance of the physical principles which produce or influence the phenomena they study. Another approach, called *deterministic,* studies the hydrologic system as a whole and attempts to model all the various mechanisms and their interactions. Empirical and statistical studies have provided many of the formulations or submodels used in deterministic models. These models, of which several are in use, take the form of digital computer programs (Crawford & Linsley, 1966; James, 1970; Holtan & Lopez, 1971). A large group of them were inspired by the Stanford Model (Crawford & Linsley, 1962). The mechanism submodels are generally written as subroutines called by the main program and may easily be modified or completely replaced when improved mathematical representations of mechanisms are found.

Another deterministic approach, *hydrologic physics,* applies the fundamental laws of physics, insofar as they are understood, to various physical mechanisms. This approach is still in its infancy and has as yet produced no hydrologic models of consequence although its philosophy has been discussed (Freeze & Harlan, 1969). It is an approach in which soil scientists can make a very real contribution and will be discussed in greater detail in a later section.

The five major approaches just discussed fulfill different needs and, therefore, do not conflict or compete with each other. Empirical hydrology and hydrologic physics complement each other and are essentially research approaches. Their aim is to discover and characterize hydrologic processes. Empiricism is often active in the discovery process and in first approximations to characterization. Physics refines and extends the characterization of mechanisms and also leads to useful discoveries. Statistical, stochastic, and the nonphysically based deterministic approaches have research missions, but probably find their greatest use as predictive tools in applied hydrology.

Hydrologists have been concerned mainly with bulk water considerations such as flood peaks, minimum low flows to be expected, *in situ* quality, etc. Statistical, stochastic, deterministic, and many empirical studies are pursued with these types of objectives in mind. There is still room for improvement in these traditional hydrologic efforts. Streamflow from ungaged areas may often be predicted with fair accuracy, but there are also many disappointments. Prediction of the hydrologic effects of land use changes is very risky indeed. These are what might be called traditional hydrologic problems. There are others of more recent origin.

Now that the populace has awakened to environmental problems, hydrologists see renewed relevance in the fact that water is one of the prime movers of substances in nature, either in solution, in suspension, or as a moving bed load. Present surface water approaches combined with erosion and sediment control practices can probably cope with problems involving suspended and bed load materials because such movements take place in connection with surface water flow. However, dissolved substances move with water wherever it goes. Furthermore, these substances are subject to physical, chemical, and biological modification as they move through different environments or simply with the passage of time.

The investigation of water-carried substances is an area where hydrologists, soil scientists, and agronomists will no doubt complement each other's efforts. Hydrologists should be able to provide information on water flow paths and velocities in any part of the ecosystem. Soil chemists and microbiologists should be able to combine that information with the special knowledge and skills of their disciplines to build models of chemical reaction and microbial processes for the soil water part of the system. From these models, there may be feedback information to improve the water flow models. Hydrologists will certainly need the help of microclimatologists and plant physiologists in learning how water moves to plant roots; this is an essential need for improving present concepts and the present crude models of soil water flow. Another equally essential need is to improve our concepts of how water enters the real soil water system. Soil morphologists may have insights on the influence upon water entry of soil structure, root holes, root channels, etc.

HYDROLOGIC SYSTEMS

Physics-Based Concept

Understanding the movement of dissolved substances through the ecosystem as well as improving hydrologic modeling requires a more general and complete concept of hydrologic systems than is offered by current ideas regarding surface watersheds and ground water aquifers. These two entities, in fact, would form areas or regions of the more general system.

A complete hydrologic system may be conceived of as a three-dimensional body of porous materials. An upper surface zone supports life and is in turn hydrologically influenced by life forms. It is across or on this surface that most of the exchange of water between the hydrologic system and its environs, the atmosphere and downslope or downstream land areas, takes place. However, subsurface hydraulics is the initial concern of several investigators working with hydrologic system physics because it is seen as the cornerstone on which to build an understanding of all the other physical processes of terrestrial hydrology. Such processes as base flow, interflow, and in large part surface runoff production may simply be manifestations of subsurface hydraulics that have somehow been recognized as separate phenomena and assigned names.

The fundamental concept regarding the flow of water in porous media is Darcy's law (Darcy, 1856)

$$v = -K(dH/d\ell) \tag{1}$$

which states that water moves with velocity, v, (per unit cross sectional area, dimensions LT^{-1}) in the direction, ℓ (L), of decreasing hydraulic head, H, stated as a depth of water, (L). The proportionality factor, K (LT^{-1}), assumed constant for a given saturated soil, is called hydraulic conductivity. Its value is dependent upon both soil and water physical properties. For soil water flow, hydraulic head is considered to be the sum of pressure head, h, stated as an equivalent head of water, (L), and elevation head, Z, stated as height above an arbitrary datum, (L),

$$H = h + Z. \tag{2}$$

Darcy's law, originally stated for saturated, one-dimensional flow in porous media, has been generalized to all three dimensions and has been found to apply to unsaturated flow (Hubbert, 1956; Childs, 1969; Hillel, 1971). In the latter case, hydraulic conductivity, K, is not a constant, but varies with water content. Darcy's law can be combined with the continuity equation, which expresses the principle of the conservation of mass, and, assuming water to be incompressible, the Richards equation for water flow in soil results (Richards, 1931)

$$C(h)\frac{\partial h}{\partial t} = \frac{\partial}{\partial x}\left[K_x(h)\frac{\partial h}{\partial x}\right] + \frac{\partial}{\partial y}\left[K_y(h)\frac{\partial h}{\partial y}\right]$$
$$+ \frac{\partial}{\partial z}\left[K_z(h)\frac{\partial h}{\partial z}\right] + \frac{\partial K_z(h)}{\partial z}. \tag{3}$$

In this equation, z is vertical and positive upward. Water capacity, C (L^{-1}), is the slope of the water retention curve:

$$C = d\theta/dh \tag{4}$$

where θ is volumetric water content, $(L^3 L^{-3})$. The letters x, y, and z refer to the directions of the Cartesian axes and to distances in these directions (all of dimension L).

The Richards equation, a parabolic partial differential equation, is central to any porous media flow system. However, it is not a mathematically complete model of such a system (Smith, 1965). The model can only be completed by specifying two more classes of quantities. One is initial condition, i.e., the distribution of h throughout the system when operation of the model begins. The other is the condition, constant or variable, of the several boundaries of the system. If a boundary condition is variable, then its manner of variation in time and space must be given. The relationships $C(h)$, $K_x(h)$, $K_y(h)$, and $K_z(h)$ must also be known for the media involved.

The Richards equation and its requirements regarding boundary and initial conditions imply that a hydrologic system is not completely defined until all boundaries and their conditions are defined. The nature of the equation as evidenced by study of its solutions (Rubin, 1968; Amerman, 1969; Freeze, 1971) shows that the boundaries, though they control the system, are also influenced by it. Therefore, conditions of the upper surface boundary of a hydrologic system hydraulically reflect the geometry, physical makeup, and hydraulic condition of the entire system. Investigations of surface phenomena without regard to subsurface influence are unlikely to lead to comprehensive or rigorous understanding of hydrologic systems. This point has been largely overlooked or considered only superficially by many investigators of watershed hydrology.

For example, the familiar infiltration curve and various equations for infiltration reflect some subsurface hydraulic influences, namely the increase in hydraulic conductivity and the decrease in near-surface hydraulic gradient as a soil wets. The infiltration curve, from the nature of its measurement (Musgrave, 1935; Bertrand & Parr, 1961), applies to one-dimensional infiltration, and does not reflect the influence of the rest of the three-dimensional system upon near-surface hydraulic gradients. The same is true for infiltration equations (Horton, 1935; Philip, 1956; Holtan, 1961) and for many infiltration models (Hanks & Bowers, 1962; Whisler & Klute, 1965).

Figure 1 illustrates some of the concepts just discussed. The lateral boundaries of surface watersheds are topographic divides. There is no flow from one side of the divide to the other. Mathematically, on the divide the hydraulic energy gradient, dH/dn, normal to the vertical plane of the divide is zero. In isotropic materials, subsurface boundaries, or divides, are defined by the locus of points or an imaginary surface over which the normal energy

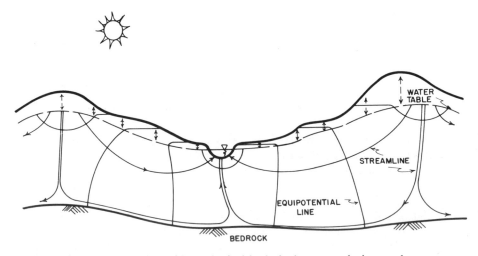

Figure 1. Cross section of hypothetical hydrologic system during a relatively dry period.

gradient is zero. The subsurface dividing boundary running through permeable material need not be vertical. It may even shift about, within certain limits, under the influence of different precipitation patterns. The hydraulic energy gradient normal to impermeable boundaries is also zero.

The solution of equation [3] with the required boundary and initial conditions yields, for any given time, the distribution of pressure head, h, within the hydrologic system (Amerman, 1969). By applying equation [2] to this distribution, surfaces can be determined on which H is everywhere equal. These are called equipotential surfaces (equipotential lines in two dimensions), and the direction and velocity of flow can be approximated by applying Darcy's law to the portions of the media between them. In isotropic media, direction of flow is perpendicular to the equipotential surfaces and in the direction of decreasing H. In two dimensions, stream surfaces are portrayed as streamlines. The streamlines are also pathlines for steady flow systems. However, most hydrologic systems are either wetting or drying, so that the streamlines are continually shifting. The path followed by a molecule of water can be approximated from chronological plotting of streamlines and calculation of velocities.

From the h-distribution itself, isobars (surfaces of equal pressure head) may be determined. The most important of these is the zero isobar, called the water table. Above the water table, h-values are negative; below it, they are positive. Usually, the water table is considered to divide the saturated zone below it from that which is unsaturated above it. However, this is not necessarily true because air may be trapped below the water table and will not enter soil pores until the pressure head falls below a certain value, the air entry value, which may be significantly negative in some soils.

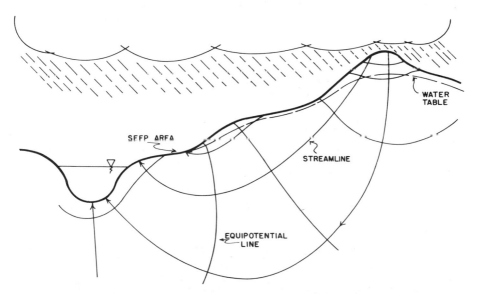

Figure 2. Section of hypothetical hydrologic system during a wet period.

In the unsaturated zone, hydraulic conductivity, K, diminishes as water content, θ, decreases. This happens as a result of the emptying of some pores and the consequent loss of part of the flow path together with an increase in tortuosity of the remaining flow path. Pressure head, h, (negative in the unsaturated zone) also decreases with decreasing water content. Between the saturation and wilting point values of θ, both K and h usually vary over several orders of magnitude. Under some conditions, water content may vary rapidly over a short distance, in which case the influence of gravity on the hydraulic gradient will be overshadowed by the influence of pressure head difference. Flow may then occur in any direction. The double-headed arrows in the upper part of the system portrayed in Fig. 1 illustrate that flow direction may have an upward component near the upper surface and a downward component near the water table. Their orientation also illustrates that during dry periods, flow in the unsaturated zone may become essentially one-dimensional, so that when precipitation begins, initial infiltration is one-dimensional and downward over the entire watershed.

Figure 2 is an expansion of the right-hand portion of Fig. 1 during a very wet period, perhaps in the spring. The water table is higher, and in this hypothetical hydrologic system, it has intersected the soil surface near the stream. The soil above the water table is very wet. Under such conditions, hydraulic gradients in the unsaturated zone are no longer one-dimensional. Their vertical component has decreased to a greater extent near the bottom of the slope than it has at the top, and the infiltration rate will drop off more rapidly low on the slope than it does near the ridge top. Below the point where the water table intersects the surface, hydraulic gradients have vertically upward components and subsurface water returns to the surface, a reversal of the infiltration process. From this it should be clear that one-dimensional infiltration curves and equations should be used with caution and may not be applicable at all in some situations.

Completing the Hydrologic System Concept

The porous media considerations of the previous section laid a cornerstone for hydrologic system concepts, but by no means can they erect the entire structure. The superstructure of the hydrologic system concept will apparently depend upon developments in surface water hydraulics and in soil-plant-water-air investigations. Work is proceeding in both these areas. Equations similar to the Richards equation have been developed for overland flow (Liggett & Woolhiser, 1967), and at least one technique for simultaneously considering overland flow and infiltration has been proposed (Smith & Woolhiser, 1971).

Soil scientists have been investigating the movement of water through plants to the atmosphere; physically based models of the process have recently been proposed (Stewart & Lemon, 1969; Molz & Remson, 1970). How to model the movement of water from soil into and through the plant is not yet fully understood. Thus, the upper surface boundary condition for solution of the Richards equation cannot yet be adequately stated for vegetated systems.

The occurrence of flora and fauna in the upper part of hydrologic systems probably will lead to a boundary concept different from that usually associated with the Richards equation and others of its type. These boundaries are surfaces or lines without thickness upon which pressure heads or energy gradients can be defined. The upper surface of the porous media part of vegetated hydrologic systems is hydraulically diffuse, i.e., water may enter and leave the media not only across the visible air-soil interface but also through structural entities such as interpedal facies, worm holes, root channels, etc.

A completely satisfactory physics-based model for hydrologic systems is not expected in the near future, but progress is being made. Scientists from many disciplines, particularly soil science and agronomy, can contribute much toward attaining this goal.

Use of the Hydrologic System Concept

The hydrologic system concept, as it has been developed to date, using porous media flow theory, gives insight into several significant questions in watershed hydrology such as partial area runoff production and the applicability of infiltrometer measurements under rainfall conditions. It has indicated why some small watershed field investigations have been unsuccessful in giving information on the hydrology of larger watersheds. It can be expected that further development of the concept and expansion to include surface runoff and evapotranspiration will shed light on other problems of significance.

The hydrologic system concept may provide background understanding for judging the worth of and improving some of the empirically-based submodels of certain deterministic models. It may lead to the development of some physics-based submodels which can be substituted for empirical ones. It is even conceivable that entirely physics-based models can be developed as alternatives to other types of deterministic models.

Physics-based models can potentially be used for many purposes besides precipitation-runoff investigations. Using such models, water availability at any point in a hydrologic system, including vegetation root zones, can be the subject of study and prediction, as can the movement of subsurface waters and the dissolved substances carried with them.

Rigorous, detailed models, based on equations of the Richards type, would be complex, digital computer models probably requiring machines of larger storage capacity than are available today. Many measurements would be required of several physical characteristics of any real hydrologic system to which they would be applied. Costs might seem to be prohibitive; yet, acceptability of cost is a relative matter and must be judged against needs and the value of returns. In certain circumstances involving human health, there may be no alternatives to a physics-based hydrologic system model or an approximation to it.

Rigorous models may not be necessary in all circumstances. Rather involved computer models for one-dimensional infiltration are available today (Hanks & Bowers, 1962; Whisler & Klute, 1965) and are used to study the process and to predict infiltration for complex situations. However, for certain situations, the Richards equation can be simplified or approximated by simple algebraic forms (Philip, 1956). These so-called analytic models are relatively easy and inexpensive to operate. It might be expected that some, perhaps many, hydrologic system geometries, properties, and inputs can be approximated by analytic models based on a physical understanding of hydrologic systems.

SOILS AND HYDROLOGY

Soil Physics

The preceding sections have been devoted to discussion of hydrologic concepts. References to soils were absent in discussing stochastic hydrology, present in discussing deterministic models, and were all-pervasive in consideration of hydrologic physics and the development of hydrologic system concepts. Soil physics has contributed immensely to porous media flow theory used to establish a cornerstone for the system concept. It is likely that soil physics and other subdisciplines of soil science will be a primary source in completing the foundation and in building the part of the superstructure concerned with evapotranspiration. Air movement and entrapment within subsurface regions and water flow laws for swelling soils are areas in which soil physics is probably in a good position to lead in research. These are areas that must be researched for the improvement of physics-based hydrologic system models. Soil physicists and plant physiologists are probably also in a good position to find the linkage mechanism between evapotranspiration and the porous media flow region.

Soil Survey and Morphology

We have briefly discussed a hydrologic systems concept that has the potential for allowing analysis and modeling of hydrologic systems *in toto*. Most of the development to date has been theoretical—the development of equations, the programming of simple preliminary models on computers, etc; and as noted earlier, theoretical development is still incomplete.

Even if physics-based models were fully developed in all their theoretical detail, we would find it very difficult to apply them to real situations in the field because we have little applicable field data.

The types of information that will be necessary include (i) system boundaries, (ii) identification and three-dimensional bounding of hydraulically homogeneous subunits within the system, and (iii) hydraulic properties of the subunits.

Mapping techniques already known to soil surveyors may find useful application to some of these needs. The surface boundaries of a system will offer no problem. Mapping subsurface boundaries may simply mean the development of isopach maps in some cases or of mapping the water table in others. In some situations, it will be necessary to measure hydraulic gradients, perhaps under a variety of conditions. Is it possible that there are soil features other than those shown by isopachs which can be used to determine the direction of water flow and thus indicate the position of a water divide?

Hydraulic properties to be measured include soil water pressure head as a function of water content and hydraulic conductivity as a function of water content or of pressure head. The first is already routinely measured.

Soil physicists have long been searching for a quick, easy method of measuring hydraulic conductivity as a function of pressure head in unsaturated soil. Several methods have been developed for the laboratory, but there is little agreement as to which is best. A recently reported *in situ* technique for measurement in fairly wet soils (Bouma et al., 1971) gives us some hope. This procedure together with one of the more promising laboratory methods for the drier range of water content may be found satisfactory for survey purposes.

However, there may be other problems. Physics-based developments so far have evolved around "glass bead" type soils. We have previously mentioned the influence of soil cracks, worm holes, roots, and root holes with respect to the surface boundary condition. What of the influence of soil structure and other soil characteristics upon water flow within a body of soil? Flow paths around soil peds probably work in conjunction with flow paths within the peds. Under wet conditions, the influence of the former may predominate, but under dry conditions the latter may more significantly influence soil water movement. Bouma and Anderson (1973, Chap. 5, this book) have considered this question in some detail. Since some peds may become quite large, large samples may be necessary for the determination of hydraulic conductivity in the wet range of water content. Smaller samples may suffice in the drier range. The last few sentences have been speculative, but illustrate that soil surveyors, with their intimate knowledge of the morphology of soils, may have insights which will be very useful in studying the hydraulic properties of soils.

Identification and mapping of hydraulically homogeneous subunits in a hydrologic system will depend in part upon finding suitable techniques for measuring hydraulic conductivity. However, it may be that hydraulic properties of soils will be found to correlate with certain morphological properties (Bouma & Anderson, 1973, Chap. 5, this book). Mapping of subunits with depth will be necessary as well as mapping laterally over a surface.

Definition of what constitutes a hydraulically homogeneous subunit will certainly be a subject for discussion and study. What is the optimum range in a given characteristic for a subunit? Will a subunit under wet conditions continue to be acceptably homogeneous under dry conditions, or vice versa?

The above is related to the question of the precision to which mapping and measurements must be made. Accuracy requirements will be established by sensitivity analyses of the models as they develop, together with knowledge of the allowable error limits on management or scientific information to be obtained from the models. The number of measurements necessary to achieve the needed accuracy in specifying the hydraulic properties of a subunit will then be a statistical problem.

CONCLUSION

Consideration of the needs and direction of both scientific and applied hydrology indicates that an improved understanding of the hydrology of soils must be gained and used. Hydrologists are beginning to see this need and are looking to soil scientists for assistance in gaining the necessary understanding. Because there is a need and because they are in a good position to approach soil water problems, some soil physicists, soil surveyors, and soil morphologists may wish to familiarize themselves with hydrologic viewpoints and orient at least part of their effort in that direction.

Other soil scientists and agronomists, including soil chemists, microbiologists, microclimatologists, plant physiologists, and soil fertility specialists may also wish to investigate hydrology both as a source of information for their investigations and as a field in need of contributions from their fields of interest.

LITERATURE CITED

Amerman, C. R. 1969. Finite difference solutions of unsteady two-dimensional, partially saturated, porous media flow. Ph.D. Thesis. Purdue University. Univ. Microfilms. Ann Arbor, Mich. (Diss. Abstr. 30:4105B).

Barnes, B. S. 1939. Structure of discharge recession curves. Trans. Amer. Geophys. Union 20:721-725.

Beard, L. R. 1967. Simulation of daily streamflow. Proc. Int. Hydrol. Symp. (Ft. Collins, Colo.) 1:624-632.

Bertrand, A. R., and J. F. Parr. 1961. Design and operation of the Purdue sprinkling infiltrometer. Purdue Univ. Res. Bull. 723. 16 p.

Biswas, A. K. 1965. The hydrologic cycle. Civil Eng. 34:70-74.

Bouma, J., and J. L. Anderson. 1973. Relationships between soil structure characteristics and hydraulic conductivity. Chap. 5 (this book). *In* R. R. Bruce et al. (ed.) Field soil water regime. SSSA Spec. Publ. no. 5. Soil Sci. Soc. Amer., Madison, Wis.

Bouma, J., D. I. Hillel, F. D. Hole, and C. R. Amerman. 1971. Field measurement of unsaturated hydraulic conductivity by infiltration through artificial crusts. Soil Sci. Soc. Amer. Proc. 35:362-364.

Childs, E. C. 1969. An introduction to the physical basis of soil water phenomena. John Wiley & Sons, Ltd., London. 493 p.

Chow, Ven Te, and S. Ramaseshan. 1965. Sequential generation of rainfall and runoff data. Amer. Soc. Civil Eng., J. Hydraul. Div. HY4:205–223.

Committee on Surface Water Hydrology. 1965. Parametric hydrology and stochastic hydrology. Amer. Soc. Civil Eng., J. Hydraul. Div. HY1:119–122.

Crawford, N. H., and R. K. Linsley. 1962. The synthesis of continuous streamflow hydrographs on a digital computer. Stanford Univ., Dep. Civil Eng., Tech. Rep. No. 12. 121 p.

Crawford, N. H., and R. K. Linsley. 1966. Digital simulation in hydrology: Stanford watershed model IV. Stanford Univ., Dep. Civil Eng., Tech. Rep. No. 39. 210 p.

Darcy, H. 1856. Les fontaines Publiques de la Ville de Dijon. V. Dalmont, Paris. 647 p.

Fiering, M. B., and B. B. Jackson. 1971. Synthetic streamflows. Water Resour. Monogr. 1, Amer. Geophys. Union. 98 p.

Ford, P. M. 1959. Multiple correlation in forecasting seasonal runoff. U. S. Dep. Interior, Bur. Reclam. Eng., Monogr. No. 2. 41 p.

Freeze, R. A. 1971. Three-dimensional, transient, saturated-unsaturated flow in a ground-water basin. Water Resour. Res. 7:347–366.

Freeze, R. A., and R. L. Harlan. 1969. Blueprint for a physically-based, digitally-simulated hydrologic response model. J. Hydrol. 9:237–258.

Grace, R. A., and P. S. Eagleson. 1967. A model for generating synthetic sequences of short-time-interval rainfall depths. Proc. Int. Hydrol. Symp. (Ft. Collins, Colo.) 1:268–276.

Hanks, R. J., and S. A. Bowers. 1962. Numerical solution of the moisture flow equation for infiltration into layered soils. Soil Sci. Soc. Amer. Proc. 26:530–534.

Harrold, L. L., D. L. Brakensiek, J. L. McGuinness, C. R. Amerman, and F. R. Dreibelbis. 1962. Influence of land use and treatment on the hydrology of small watersheds at Coshocton, Ohio, 1938–1957. U. S. Dep. Agr. Tech. Bull. 1256. 194 p.

Hillel, D. 1971. Soil and water, physical principles and processes. Academic Press, New York. 288 p.

Holtan, H. N. 1961. A concept for infiltration estimates in watershed engineering. U. S. Dep. Agr., Agricultural Research Service. ARS 41–51. 25 p.

Holtan, H. N., and N. C. Lopez. 1971. USDAHL-70 model of watershed hydrology. U. S. Dep. Agr. Tech. Bull. 1435. 84 p.

Horton, R. E. 1935. Surface runoff phenomena, analysis of the hydrograph. Horton Hydrol. Lab. Publ. 101. 73 p.

Hubbert, M. K. 1956. Darcy's law and the field equations of the flow of underground fluids. J. Petrol. Technol. 8:222–239.

James, L. D. 1970. An evaluation of relationships between streamflow patterns and watershed characteristics through the use of OPSET. A self calibrating version of the Stanford watershed model. Univ. Kentucky Water Resour. Inst. Res. Rep. No. 36. 117 p.

Liggett, J. A., and D. A. Woolhiser. 1967. Finite-difference solutions of the shallow water equations. Amer. Soc. Civil Eng., J. Eng. Mech. Div. EM2:39–71.

Lowdermilk, W. D. 1926. Forest destruction and slope denudation in the Province of Shansi. China J. 4:127–135.

Matalas, N. C. 1967. Mathematical assessment of synthetic hydrology. Water Resour. Res. 3:937–945.

Molz, F. J., and I. Remson. 1970. Extraction term models of soil moisture use by transpiring plants. Water Resour. Res. 6:1346–1356.

Musgrave, G. W. 1935. The infiltration capacity of soils in relation to the control of surface runoff and erosion. J. Amer. Soc. Agron. 27:336–345.

Mustonen, S. E. 1967. Effects of climatologic and basin characteristics on annual runoff. Water Resour. Res. 3:123–130.

Philip, J. R. 1956. The theory of infiltration: 4. Sorptivity and algebraic infiltration equations. Soil Sci. 84:257–264.

Richards, L. A. 1931. Capillary conduction of liquids through porous mediums. Physics 1:318–333.

Rubin, J. 1968. Theoretical analysis of two-dimensional, transient flow of water in unsaturated and partly unsaturated soils. Soil Sci. Soc. Amer. Proc. 32:607–615.

Schreiber, H. A., and D. R. Kincaid. 1967. Regression models for predicting on-site runoff from short-duration convective storms. Water Resour. Res. 3:389–395.

Sherman, L. K. 1932. Streamflow from rainfall by unit-graph method. Eng. News-Rec. 108:501–505.

Smith, G. D. 1965. Numerical solution of partial differential equations. Oxford Univ. Press, New York. 179 p.

Smith, R. E., and D. A. Woolhiser. 1971. Mathematical simulation of infiltrating watersheds. Colorado State Univ. Hydrol. Paper No. 47. 44 p.

Snyder, W. M. 1962. Some possibilities for multivariate analysis in hydrologic studies. J. Geophys. Res. 67:721–729

Stewart, D. W., and E. R. Lemon. 1969. The energy budget at the earth's surface: A simulation of net photosynthesis of field corn. U. S. Army Electronics Command Tech. Rep. ECOM 2-68 I-6. 132 p.

Veitch, L. G., and K. J. Sheperd. 1971. A statistical method for flow prediction, River Murray example. Water Resour. Res. 7:1469–1484.

Whisler, F. D., and A. Klute. 1965. The numerical analysis of infiltration, considering hysteresis, into a vertical soil column at equilibrium under gravity. Soil Sci. Soc. Amer. Proc. 29:489–494.

Wicht, C. L. 1965. The validity of conclusions from South African multiple watershed experiments. p. 749–760. *In* W. E. Sopper and H. W. Lull (ed.) Proc. Int. Symp. Forest Hydrol. (Penn State Univ.) Pergamon Press, Oxford, N. Y.

Wiser, E. H. 1967. Stochastic models in hydrology. Trans. Amer. Soc. Agr. Eng. 10: 813–819.

The Role of Soil Water in the Hydrologic Behavior of Upland Basins[1]

10

WADE L. NUTTER[2]

ABSTRACT

The distribution of soil water in upland basins greatly affects the extent of source areas and the response patterns of both storm and between-storm streamflow. Except during the most extreme storms, all the precipitation falling on well-vegetated slopes infiltrates and while some reappears in the channel as stormflow, a major portion of the rain remains in the basin as dynamic storage. During a storm, the stormflow source area expands out from the stream channel as slopes contribute primarily unsaturated subsurface flow and the channel system lengthens. After the storm ceases, source areas may continue to expand as subsurface flow feeds the lower slopes near the channel, often leading to a second hydrograph peak several hours or days after the rain ceases. As the channel system and source areas recede, unsaturated subsurface flow continues to sustain baseflow. Basin parameters that affect the soil water regime and associated soil water energy conditions, and therefore the distribution of source areas, are slope length from channel to ridge, angle of slope, regolith depth, and regolith physical properties. Physical models of hillslope segments have provided some insight into the interrelations among the basin parameters as well as the flow pathways and source areas of subsurface flow.

INTRODUCTION

The fact that soil water conditions may influence the hydrologic behavior of watersheds is well recognized. Yet efforts to translate this knowledge into a working model describing the various sources, pathways, mechanisms, and timing delays that underlie the dynamics of stream discharge from headwater basins have met with little success. One of the primary reasons for this failure has been the explanation of the storm runoff process almost entirely in terms of overland flow. On permeable upland slopes in humid regions, overland flow is rarely observed and subsurface flow in the vicinity of the stream channel accounts for much of the storm runoff (stormflow). Important to

[1] Contribution from the School of Forest Resources, University of Georgia, Athens. Research supported in part by the Office of Water Resources Research, U. S. Department of Interior, as authorized under the Water Resources Research Act of 1964, P. L. 88–379.

[2] Associate Professor of Forest Hydrology, School of Forest Resources, University of Georgia, Athens.

the subsurface flow process, and watershed behavior in general, are the basin's physical characteristics. These characteristics will affect the distribution and flow pathways of water in the soil mantle and the timing of flow from the stormflow source areas to the channel.

To understand why so little emphasis has been placed on the role of soil water in the hydrologic behavior of basins we must first review briefly some recent history of hydrology. The work of Robert E. Horton (1945), a geo-morphologist and hydrologist in the 1930's and 1940's, placed considerable emphasis on the role of infiltration and surface soil erosion on the develop-ment of stream channel patterns by overland flow. Horton's work implied that rainfall exceeding the infiltration capacity, of rainfall excess, caused overland flow and was responsible for the immediate stormflow responses in the channel system. Following the work of Horton, a "technology of over-land flow" developed and very little emphasis was placed on the more critical "basin-wide" infiltration process. Attention was focused on overland flow as the predominant hydrologic process to the extent that all other processes were virtually ignored. Subsurface flow (sometimes referred to as interflow and quick return flow) was considered to move at velocities too slow to be-come part of the stormflow hydrograph and consequently was largely ignored as an important hydrologic process. As a result, hydrology has been "surface process" oriented and the mechanisms and pathways of water movement through the soil profile have received little attention by hydrologists.

Horton's infiltration model and subsequent flow models have been modified to a certain degree, but most still rely on overland flow as the pri-mary source of stormflow, consequently forcing virtually all subsurface flow into overland pathways. Within the past 10 years a few hydrologists have questioned the validity of the overland flow model as applicable to humid region vegetated basins. This has led to a greater emphasis placed on the movement of soil water and source areas of streamflow during both storm and between-storm periods (Hewlett & Hibbert, 1967; Kirkby & Chorley, 1967; Betson & Marius, 1969; Hewlett & Nutter, 1970; Ragan, undated). In actuality, the overland flow model is just one end member of a spectrum of possible flow models. An entirely subsurface flow model would represent the other end member. The overland flow models seem best suited to the domain of basins where rainfall intensities are very high and vegetation sparse. On the other hand, in well-vegetated basins the operational flow models would be closer to the subsurface flow end of the spectrum.

To avoid misunderstanding, overland flow is defined as rainwater that fails to infiltrate the soil surface at any point along its way from the basin surface to the stream channel. This definition is necessary to separate, for the purposes of this discussion, the rapid delivery of overland flow to the stream channel from the much slower delivery of subsurface flow. Overland flow may move at velocities up to a 8 cm/sec, more than 6,000 m/day, whereas subsurface velocities at saturation seldom exceed 4 cm/day. Rain-water that infiltrates and later seeps out only a meter downslope has already been delayed long enough at these velocities to alter the characteristics of the stormflow hydrograph. This fine line of distinction is important because whether water has infiltrated and moved only a meter before appearing as a surface extension of the channel system or whether it has not infiltrated at all has a great deal to do with the nature of a basin's hydrologic behavior.

Forest hydrologists are well aware that overland flow occurs rarely or in very small amounts from forests and many wildlands. This is not so strange to anyone who has stood on vegetated slopes during downpours and, after watching all the water enter the soil, has later seen it seep into the stream channel. In addition, the entire basin is not contributing equally to stormflow. In other words, depending on antecedent conditions and physical characteristics of the basin, various portions of a basin may respond differently to rainfall input by contributing variable quantities of stormflow to the stream channel, not only during the storm itself but for considerable periods of time following the storm. Thus, some areas of the basin contribute more stormflow than others and it is conceivable that some portions of a basin contribute stormflow only during the most extreme storms (Betson & Marius, 1969; Hewlett & Hibbert, 1967; Dickinson & Whiteley, 1970; Ragan, undated).

In conclusion, infiltration as a critical hydrologic process is frequently not a limiting factor in humid region well-vegetated basins when the total mosaic of vegetation patterns and basin shape is considered. Infiltration capacities may be momentarily exceeded during an intense burst of rain but most of this excess rainfall is ponded or infiltrates the soil and never reaches the stream channel as overland flow. Average infiltration over a basin under natural rainfall is quite different from plot infiltration capacities determined by rainfall simulators or similar techniques. Infiltration capacity curves drawn under such conditions can rarely be used to predict the infiltration excess and stormflow during natural storms.

SLOPE WATER MOVEMENT

Hoover and Hursh (1943) and Roessel (1950) first focused attention on the importance of slope water movement in the hydrology of upland forested basins. However, it was almost 20 years before Hewlett (1961b) reported the results of a hillslope drainage study that presented strong evidence that the storage and movement of soil water on slopes predominantly influenced the stormflow and baseflow characteristics of a basin. It was apparent from the hillslope study and others to follow that the important physical characteristics of a basin controlling the water and energy conditions and the rate and source of outflow from a draining slope were (i) angle and length of slope and (ii) depth and physical properties of the hydrologically active portions of the soil mantle or regolith (Hewlett & Hibbert, 1963; Whipkey, 1967; Troendle, 1970).

Occurrence in Hillslope Segments

The last of four hillslope models constructed at the U. S. Forest Service Coweeta Hydrologic Laboratory, Franklin, North Carolina, was 61 m long, 2.1 m deep, and 1.2 m wide on a natural 35% slope packed with soil to as uniform a density as possible. In essence, the model provided a nearly isotropic hillslope segment roughly half the average length of slopes in the Southern Appalachian mountains. Impermeable segment sides and bottom were formed with several layers of plastic and an outflow pipe in a headwall at the lower end permitted manipulation of water table level and continuous measurement of outflow. Overland flow was measured with an outflow pipe at the top of the wall, but after grass covered the soil there was no overland flow. At no time in over 7 years of operation did the water table emerge through the soil surface nor did drainage cease. Volumetric water contents and soil water pressures in the segment were monitored with a neutron scattering device and tensiometers, respectively.

Comparison of discharge hydrographs from the hillslope segment and the 15-ha watershed on which it was built are shown in Fig. 1 and 2. A 101-mm (3.96-inch) September 1963 storm (Fig. 1) on well-drained soils partially dried by evapotranspiration demonstrates how quickly subsurface flow can appear as outflow. The main portion of the storm ended shortly before the major stormflow peak at 16 hours, although a short burst of rainfall caused

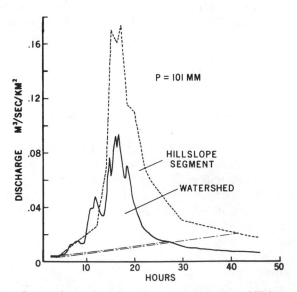

Figure 1. Comparison of outflow from the 61-m long hillslope segment and its companion 15-ha watershed following a 101-mm storm in September 1963. Stormflow is arbitrarily separated from baseflow by a straight line.

another small peak at 18 hours. The 61-m segment produced an almost simultaneous hydrograph peak twice as high as the watershed's peak and a stormflow duration half again as long. [Stormflow is arbitrarily separated from baseflow by a straight line with a slope of 5 × 10⁻⁴ m³ sec⁻¹ km⁻² hour⁻¹, (Hewlett and Hibbert, 1967)]. The segment had no active channel at any time during the stormflow period and all outflow was the result of subsurface flow. Although the rain was sufficient to produce high water contents well up the slope of the model, outflow receded rapidly.

The watershed reveals its greater depth and length of slope by storing a larger proportion of the rain and by dropping its rate of discharge even more rapidly. In terms of volume of stormflow produced, the watershed yielded 2% of the rain received and the hillslope segment yielded 5%. Included in the watershed yield is the rain falling directly into the channel, or channel precipitation. Considering the absence of overland flow and the behavior of the segment during this storm, it can be concluded that only a narrow zone, or source area, along the channel yielded subsurface flow to the channel system.

During an even larger storm of 143 mm (5.63 inches) in September 1965, (Fig. 2) the watershed yielded a substantial peak, stored 94% of the rain, and dropped rapidly back to normal flow. Most of the storm's rainfall occurred during two sustained periods, corresponding to the two peaks in watershed outflow. No rain occurred after the last peak. Again only a small portion of the total watershed area, primarily the near-channel areas, yielded stormflow. The segment produced about the same peak flow as before (perhaps near the upper limit of the segment to produce water) and although stormflow continued for 5 days at a well-sustained rate, about 80% of the rain remained in the segment after stormflow ended. Again, there was no

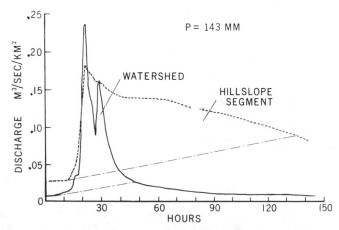

Figure 2. Comparison of outflow from the 61-m long hillslope segment and its companion 15-ha watershed following a 143-mm storm in September 1965. Stormflow is arbitrarily separated from baseflow by a straight line.

overland flow or an active channel in the segment at any time during the stormflow period.

As water was fed to the water table during the storm, not only from soil directly above but also laterally from upslope, the water table rose and the effective storage depth at the lower end of the segment was reduced. However, storage was still adequate because the downslope subsurface travel time served as a form of dynamic storage. The surprisingly well-sustained rate of outflow was the result of (i) the water table receding and (ii) unsaturated flow from the lower two-thirds of the slope. In essence, the stormflow source area expanded upslope during and for a short time after the storm and then slowly receded downslope during the period of sustained outflow. The watershed again exhibited its greater depth, on the average two to three times the segment's depth, and its greater slope length by storing more of the rainfall. It required a 500-mm (20-inch) rainstorm of 3 day's duration to force the watershed to deliver 20% of the rainfall as stormflow and to behave as the segment did in this 143-mm storm.

The previous two examples illustrate slope water movement during and following rainfall. Consider now the gradual redistribution of water in a hillslope segment with evapotranspiration prevented. An earlier hillslope segment built at Coweeta, North Carolina and reported by Hewlett and Hibbert

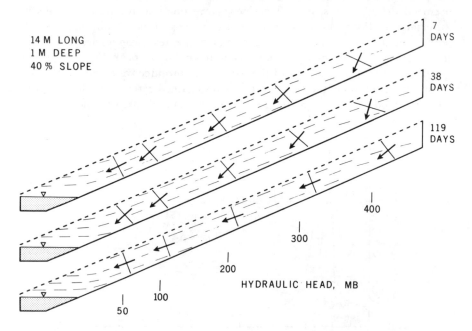

Figure 3. A drainage sequence of the covered 14-m long hillslope segment illustrating the reorientation and direction of flow lines during drainage. The dashed lines represent isolines of matric suction measured at four vertical banks of tensiometers. The first dashed line above the water table represents −25 mbars and each additional line represents an increment of −25 mbars. Flow directions are shown by the arrows.

(1963) was 0.9 m deep, 0.9 m wide, and 13.7 m long on a 40% slope. The segment consisted of a concrete trough packed with a sandy loam soil to a relatively uniform bulk density. After construction, the segment was thoroughly soaked and covered to prevent evaporation and further wetting by rainfall. A sequence of soil water energy conditions within the segment during drainage is shown in Fig. 3. Isolines of matric suction and hydraulic head and the resulting flow lines are shown at intervals of 7, 38, and 119 days after drainage began. The average water content after soaking was approximately 40% by volume and at the end of the drainage cycle was approximately 34%.

At the beginning of the drainage cycle when the soil mass was wet, the actual flow direction was oriented towards the bottom, although the net flow was downslope. As drainage continued the flow lines slowly oriented parallel with the surface. Whisler (1969) reported a similar flow pattern for a theoretical analog model under steady-state flow conditions. Similar energy and flow-line patterns have been observed in the laboratory on a variable slope model that is 6 m long, 1.5 m deep, and 0.2 m wide. This segment model is designed to study the interrelationships among various physical characteristics of slopes, the physics of unsaturated water movement in large soil masses, and problems associated with scaling flow models.

The preceding example of a covered hillslope segment perhaps best illustrates the drainage of deep subsoil horizons where there is little withdrawal of water by plant roots. Figure 4 shows how the energy gradients and flow lines would appear in relatively isotropic surface horizons during a period of evapotranspiration and rainfall. The energy gradients and flow lines

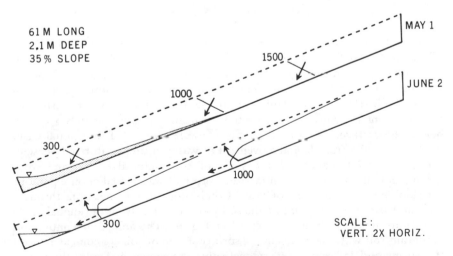

Figure 4. A drainage sequence on the grass-covered 61-m long hillslope segment illustrating the reorientation and direction of flow lines during a period of drainage and evapotranspiration. The numbers represent the hydraulic head in mbars. Flow directions are shown by the arrows.

in the 61-m long hillslope segment at Coweeta, North Carolina, are shown for the period 1 May, following a soaking rain, to 2 June while the segment was in grass cover.   At the beginning of the period the flow lines resemble the initial wet stage of Fig. 3.   However, as drainage continues and the surface meter of soil is dried by evapotranspiration, the general flow pattern remains laterally downslope, although a considerable portion of the water is directed to the surface under energy gradients created by evapotranspiration.   At no time during this 1-month period, or for the remainder of the growing season, did outflow cease.   We could expect similar conditions to exist on natural hillslopes leading to dry upper slopes and wet lower slopes as water slowly migrates downslope.   Thus, in the advent of a storm, the lower slopes will re-spond more quickly and produce outflow rapidly enough to become storm-flow.   The mechanics of stormflow production may be viewed as a relaxation of the energy gradients upslope as the storm continues.   The response of this relaxation will vary with rainfall intensities, antecedent conditions, and basin physical characteristics.

Although these results were obtained for hillslope segments with rela-tively isotropic soil and an impermeable artificial bottom, there is evidence that similar lateral flow conditions exist on natural slopes.   Minshall and Jamison (1965) report similar results on slopes with claypan soils.   Zaslavsky and Rogowski (1969) present a general model on how subsurface movement of water might contribute to the formation of soil horizons on slopes.   They conclude that the magnitude and direction of soil water movement will de-pend on the degree of anisotropy and slope of the soil mass.

Occurrence in Watersheds

Field evidence of the magnitude of lateral soil water movement over a 24-ha Whitehall Forest Experimental Watershed in the Georgia Piedmont has been presented by Tischendorf (W. G. Tischendorf, 1969. Tracing stormflow to varying source areas in a small, forested watershed in the Southeastern Pied-mont. Ph.D. Dissertation. University of Georgia, Athens). Characteristic of the Piedmont is a horizon of restricted permeability approximately 1 to 2 m below the surface.   As a result, a considerable portion of the annual basin discharge results from lateral movement of water in the unsaturated phase in the upper soil horizons.   Three years of soil water data gathered from throughout the basin were summarized into soil water prediction equations for slope position and time of year.   Combining all slope positions through-out the basin, the departure of the soil water content from the annual mean at various depths to 6 m is presented in Fig. 5.   The annual maximum and minimum soil water contents in the surface 0.5 m of soil are coincident with the spring and fall equinoxes, respectively.   As shown in Fig. 5, the yearly maximum and minimum water contents at greater depths are both lagged and damped with increasing depth.   At 6 m there is little change in annual soil water content.   Collaborating this evidence are the relatively minor

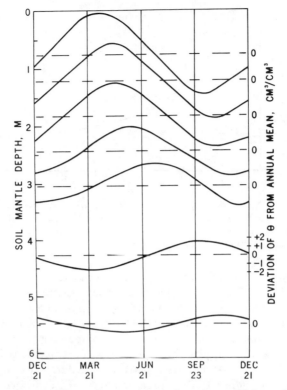

Figure 5. The deviation of water content by volume (Θ) from the annual mean as a function of depth for the Whitehall Forest Experimental Watershed in the Georgia Piedmont.

seasonal ground-water fluctuations, indicating a rather constant recharge to and discharge from the ground-water aquifers.

The main ground-water table in the Whitehall basin occurs at a depth of 10 to 15 m. Over 5 years of streamflow data indicate that baseflow from the saturated zones is rather constant throughout the year, averaging about 20% of the total annual streamflow. In the absence of overland flow, much of the remaining 80% of annual streamflow is the result of subsurface flow downslope to the channel system. Of this 80%, half leaves the basin as stormflow and half as baseflow. When a diagram similar to Fig. 5 is drawn for each of the major slope positions, lower, mid, and upper slope, a somewhat more rapid penetration of the annual recharge wave is evident for the lower slope positions than for the upper slopes. This result is to be expected because of the additional water moving laterally downslope and collecting in the lower slope positions. During the year, there is a relatively constant rate of flow through the impeding horizon into deeper soil while most of the water from large rainstorms is diverted laterally downslope.

The preceding examples show the importance of water movement in the soil mantle, particularly the surface horizons, to the hydrologic behavior

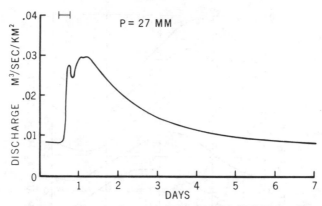

Figure 6. A stormflow hydrograph from the Whitehall Forest Experimental Watershed illustrating the delayed delivery of subsurface flow to the stream channel system following a 27-mm storm. The bar above the hydrograph represents the storm duration.

of entire basins. The soil mantle can be viewed as a zone of storage feeding baseflow and stormflow to the stream channel system. The source area for stormflow, and baseflow as well, in permeable upland basins is not a uniform or constant area, i.e., all portions of the basin are not contributing equally to streamflow. Actually combining soil water, slope energy gradients, rainfall characteristics, and basin physical parameters to predict a streamflow hydrograph is a difficult task and attempts to do so have not been successful. However, clear evidence for the predominance of subsurface movement of water in the unsaturated phase can be observed in the field under favorable meteorological conditions.

Figure 6 shows a stormflow hydrograph resulting from a brief winter rainstorm of 27 mm (1.07 inches) on the Whitehall Forest Experimental Watershed. The initial peak, which occurred during the storm, represents between 1 and 2% of the rainfall volume, actually little more than the rain that fell directly on the stream as channel precipitation. From some indeterminate area surrounding the expanding stream channel, the slowly moving subsurface water gathered to a second peak 11 hours after rainfall ceased. There is no question that this is subsurface flow; it has been verified by walking over the basin during and following storms. In addition, the maximum travel time for surface flow from the most distant part of the basin would not exceed 1 hour. As storms become larger and more complex, the hydrographs become larger and the channel precipitation peaks blend with the subsurface contributions resulting in the familiar single-peaked hydrograph. This one example from the Georgia Piedmont is not unusual. Hewlett and Nutter (1970) present evidence from several watersheds throughout the world where the second delayed peak of subsurface flow occurs as much as 4 days later.

The actual characteristics of stormflow from a watershed, really a composite of hillslope segments, depend on the pathways taken by the subsurface portions of stormflow, which in turn depend on the antecedent conditions in the soil mantle, the exact timing and intensity of the rainstorm, and the rate at which the stream channel network expands as the subsurface flow collects at the base of slopes.

A CONCEPT OF STREAMFLOW PRODUCTION

It is evident from the research conducted on hillslope segments and from field observations that the zone often referred to as the "soil moisture reservoir" has a predominant role in streamflow production from permeable upland basins. A concept has evolved over the last 10 years that accounts for soil water movement, primarily as unsaturated flow, as a fundamental process in streamflow production. This concept has been referred to as the variable, partial, and contributing source area concept (Hewlett, 1961a; Betson & Marius, 1969; Dickinson & Whiteley, 1970; Hewlett & Nutter, 1970). Although each author describes the process somewhat differently, common to all is the recognition of saturated and unsaturated subsurface flow as a source of stormflow and the nonlinear nature of stormflow production from only portions of a basin.

The hillslope segment and associated watershed research have demonstrated that long slopes can continually drain throughout the year and produce baseflow by the lateral downslope movement of water either to small zones of saturation along the stream or to the general aquifer system in proximity to the streams. Recognition of this process will help in the evaluation of soil formation and vegetation distribution and growth. For example, foresters have long recognized the improvement of site quality in headwater basins as the stream channel is approached; improved moisture relationships due to downslope migration of soil water can frequently explain the change in site quality.

Stormflow production, on the other hand, is the result of two simultaneous processes: (i) channel expansion and (ii) the subsurface delivery of water to channels. The perennial channel expands in response to subsurface flow into small draws, swampy spots, zones of low storage capacity, and intermittent channels (Gregory & Walling, 1968). The expansion is also aided by rain falling directly on these already wet areas. As the source area of subsurface flow grows, the expanding channel may grow to many times its normal width and length, often continuing to expand after the storm has ended. In essence, the stream channel system "reaches out" to tap the subsurface flow systems which can no longer transmit water beneath the surface.

The channel banks may be viewed as yielding water in a manner similar to that demonstrated theoretically by Klute, Scott, and Whisler (1965) for a saturated soil slab receiving continuous rainfall. As infiltration occurred continuously on the top half of the slab, water was flowing out of the soil surface on the lower half of the slab. Although rainfall rarely continues long enough on natural watersheds to produce saturated conditions in the surface horizons very far upslope from a channel, it is reasonable to assume that near-saturated conditions can exist at the base of a hillslope segment and that these conditions can expand upslope in a manner to suggest expansion of the channel system.

In conclusion, basin stormflow production may be viewed as a dynamic network of channels carrying off the outflow from a series of hillslope segments. The outflow source area is also a dynamic zone and for the majority of the annual stormflow represents a small area of the basin, perhaps rarely exceeding 25% of the total area in permeable upland basins. As the source area network recedes after a storm, the channel shrinks back to its original dimensions. The rate of channel expansion and shrinkage will depend on the physical characteristics of the basin, antecedent soil water conditions, and storm characteristics.

SUMMARY

The hydrologic behavior of basins is judged by characteristics of the streamflow hydrograph which represents the integration of many physical, meteorologic, and hydrologic parameters and processes. However, examination of the hydrograph alone does not reveal the sources, timing delays, or actual processes that combine to produce the hydrograph. When well-vegetated upland basins are considered to produce streamflow by a process of expanding and shrinking source areas fed by the subsurface downslope movement of water, an understanding is gained of the interaction between basin physical parameters and the flow processes which deliver water to the stream channel.

ACKNOWLEDGMENT

I wish to thank the staff of the Coweeta Hydrologic Laboratory, Southeastern Forest Experiment Station, U. S. Forest Service for providing the data on the hillslope segments and Dr. Yoshinori Tsukamoto of the Tokyo University of Agriculture and Technology who summarized part of the data while a postdoctoral research associate in forest hydrology at the University of Georgia.

LITERATURE CITED

Betson, R. P., and J. B. Marius. 1969. Source areas of storm runoff. Water Resour. Res. 5:574–582.

Dickinson, W. T., and H. Whiteley. 1970. Watershed areas contributing to runoff. p. 1.12–1.26. *In* Symposium on the results of research on representative and experimental basins. Publ. No. 96, Int. Ass. Sci. Hydrol., Gentbrugge, Belgium.

Gregory, K. J., and D. E. Walling. 1968. The variation of drainage density within a catchment. Bull. Int. Ass. Sci. Hydrol. XVIe Annee(2):61–68.

Hewlett, J. D. 1961a. Watershed management. p. 61–66. *In* Annual Report for 1961. Southeastern Forest Experiment Station, U. S. Forest Service, Asheville, N. C.

Hewlett, J. D. 1961b. Soil moisture as a source of baseflow from steep mountain watersheds. Station Paper 132. Southeastern Forest Experiment Station, U. S. Forest Service, Asheville, N. C.

Hewlett, J. D., and A. R. Hibbert. 1963. Moisture and energy conditions within sloping soil mass during drainage. J. Geophys. Res. 68:1081–1087.

Hewlett, J. D., and A. R. Hibbert. 1967. Factors affecting the response of small watersheds to precipitation in humid areas. p. 275–290. *In* W. E. Sopper and H. W. Lull (ed.) International symposium on forest hydrology. Pergamon Press, Oxford.

Hewlett, J. D., and W. L. Nutter. 1970. The varying source area of streamflow from upland basins. p. 65–83. *In* Proceedings of a symposium on the interdisciplinary aspects of watershed management. Amer. Soc. Civil Eng., New York.

Hoover, M. D., and C. R. Hursh. 1943. Influence of topography and soil depth on runoff from forest land. Trans. Amer. Geophys. Union 24:692–698.

Horton, R. E. 1945. Erosional development of streams and their drainage basins: Hydrophysical approach to quantitative morphology. Geol. Soc. Amer. Bull. 56:275–370.

Kirkby, M. J., and R. J. Chorley. 1967. Throughflow, overland flow, and erosion. Bull. Int. Ass. Sci. Hydrol. XIIe Annee (3):5–21.

Klute, A., E. J. Scott, and F. D. Whisler. 1965. Steady state water flow in a saturated inclined soil slab. Water Resour. Res. 1:287–294.

Minshall, N. E., and V. C. Jamison. 1965. Interflow in claypan soils. Water Resour. Res. 1:381–390.

Ragan, R. M. (undated). Role of basin physiography on the runoff from small watersheds. Report 17. Vermont Resources Research Center, Water Resources Research, Univ. Vermont, Burlington.

Roessel, B. W. P. 1950. Hydrologic problems concerning runoff in headwater regions. Trans. Amer. Geophys. Union 31:431–442.

Troendle, C. A. 1970. Water storage, movement, and outflow from a forested slope under natural rainfall in West Virginia (Abstract). EOS, Trans. Amer. Geophys. Union 51:279.

Whisler, F. D. 1969. Analyzing steady-state flow in an inclined soil slab with an electric analog. Soil Sci. Soc. Amer. Proc. 33:19–25.

Whipkey, R. Z. 1967. Storm runoff from forested catchments by subsurface routes. p. 773–779. *In* Proceedings of the Leningrad symposium, Floods and their computation. Int. Ass. Sci. Hydrol., Gentbrugge, Belgium.

Zaslavsky, D., and A. S. Rogowski. 1969. Hydrologic and morphologic implications of anisotropy and infiltration in soil profile development. Soil Sci. Soc. Amer. Proc. 33:594–599.

Development of Soil Temperature and Soil Water Criteria for Characterizing Soil Climates in Canada[1]

11

WOLFGANG BAIER AND ALEX R. MACK[2]

ABSTRACT

Criteria were established for describing the climatic characteristics of a soil using thermal and water parameters. Indices defining these parameters were based on standard climatological data and on soil temperature measurements, but for areas without adequate measurements, they were based on estimates derived from atmospheric data. Soil temperature at the 50-cm depth was selected to provide basic data on the thermal conditions of the site in terms of daily mean temperature and accumulated heat units. Seasons and periods of the year were established to describe the thermal condition of the site for different intensities of biological activity. Cardinal values to represent seasons and periods of different biological activities were established at mean soil temperatures of 0C (32F), 5C (41F), and 15C (59F). The moisture conditions were characterized in terms of rainfall in relation to the water requirement for potential evapotranspiration (PE) over the growing season. A storage capacity of 5 cm (2 inches) of available water was used for the major regions, but other storage capacities could be used to characterize variations in soil textures (sands, loams, clays) and topography that occur within a broad region. Eight moisture subclasses subdivided the spectrum of moisture properties during the growing season (i.e., when soil temperatures were ⩾5C) to cover the range in conditions from dominantly saturated for long periods (prehumid) to severe water deficits (subarid). Provision was made for two additional classes which do not occur in Canada. A Soil Climate Map of Canada was prepared from these characterizations showing the thermal classes and moisture subclasses (scale 1:5 million). Known physiographic features of soil types, distribution of natural vegetation, and landscape were used to define the boundaries in greater detail. Some adjustments in the criteria were made to coordinate them with the FAO/UNESCO classification system for North America.

---

[1] Contribution No. 883 of the Plant Research Institute and Contribution No. 414 of the Soil Research Institute, Research Branch, Canada Department of Agriculture, Ottawa, Canada.

[2] Research Scientist (Agrometeorology), Plant Research Institute, and Research Scientist (Soil Environment), respectively, Canada Department of Agriculture, Ottawa, Canada.

INTRODUCTION

Development of a national approach to a soil climate concept began in 1968 in Canada when the Subcommittee on Soil Climate in Relation to Soil Classification and Interpretation for the Canada Soil Survey Committee recommended that geographic and soil groups should be defined with more precise climatic attributes. Previously, climatic systems traditionally described the aerial or atmospheric conditions associated with a region frequently in relation to type and content of vegetative growth but without consideration for unique aspects affecting the soil phase of the overall biosystem. It is recognized that soil climate is a function of atmospheric conditions modified by surficial aspects such as snow, water, and vegetative cover, and by internal characteristics such as water content, organic matter, and texture.

Climate implies an expected average condition, and therefore, involves probability of a certain occurrence which is in contrast to weather which describes conditions occurring at a particular point in time and place. Consequently, climatic parameters involve data from a period of years. Two conditions of the soil system were used for describing its climate physically and its relevancy for general biological suitability. Information on its thermal condition was based on daily soil temperature at the 50-cm depth, and its moisture status was related to its ability to provide water for optimum evapotranspiration.

Following the development of criteria, a soil climatic classification system was developed and a provisional map was presented to the Eighth Meeting of the Canada Soil Survey Committee in October 1970. Subsequent modifications facilitated the interphasing of the Canadian scheme into a North American soil climatic framework as basis for the FAO/UNESCO project of a soil climate map of the continent (Clayton, 1971).

The purpose of this paper is to outline the techniques and criteria which were developed for the characterization of the temperature and moisture aspect of the soil climate and used in the preparation of the Soil Climate Map of Canada and subsequently modified for the Canadian section of the Soil Climate Map of North America.

THERMAL CHARACTERIZATION ASPECT OF SOIL CLIMATE

Seasonal Periods

Soil temperature at the 50-cm (20-inch) depth was selected to provide basic temperature data for characterizing the soil climatic conditions. Diurnal variation at 50 cm generally is negligible but temperatures at this depth reflect daily mean values and short trends in weather conditions. Where there is negligible seasonal change in temperature at the 50-cm depth (permafrost), supplementary data at the 10- and 20-cm depths and, in special areas, at the 100-cm depth were also considered.

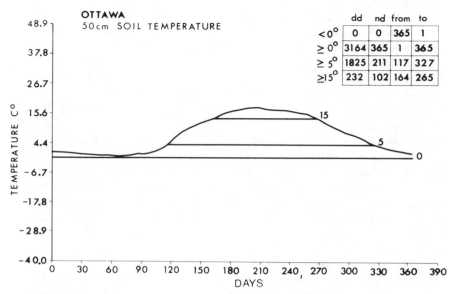

	dd	nd	from	to
$<0°$	0	0	365	1
$\geq 0°$	3164	365	1	365
$\geq 5°$	1825	211	117	327
$\geq 15°$	232	102	164	265

Figure 1. Soil temperature characteristics at the 50-cm depth. The annual cycle illustrates the length of the growing season ($\geqslant$5C), the frozen dormant period ($<$0C), and the thermal period ($\geqslant$15C). The graph obtained from computer program B121 illustrates a typical pattern of soil temperature distribution for a station (Ottawa, Can.) in Class 5.1, Mild Mesic. The insert shows the relative data on temperature characteristics for each period ($<$0C, $\geqslant$0C, $\geqslant$5C, $\geqslant$15C), namely: degree days (dd), number of days (nd), dates (from - to).

A special computer program (B121) provided listings and plottings of mean daily values interpolated from monthly temperature normals (Fig. 1). Seasons and periods were established for the duration of time of soil temperature levels associated with potential biological activities and specified thermal conditions (Table 1).

Table 1. Characterization of seasons and periods.

S_1 Dormant Season <5C (41F)

P_1—Frozen dormant period <0C (32F)

F_0 = frozen at <20-cm depth (8 inches)
F_1 = frozen at $\geq$20 cm
F_2 = frozen at $\geq$50 cm
F_3 = frozen at $\geq$100 cm

P_2—Cold dormant period $\geq$0C but <5C (41F)

F_0 = frozen at <20-cm depth (8 inches)
F_1 = frozen at $\geq$20 cm (8 inches)
F_2 = frozen at $\geq$50 cm (20 inches)
F_3 = frozen at $\geq$100 cm (40 inches)

S_2 Growing Season $\geq$5C

P_3—Cool growth period $\geq$5C but <15C

P_4—Thermal growth period $\geq$15C (59F)

    Data for these calculations were available from soil temperature records
from a 44-station network jointly administered by the Atmospheric Environment Service (AES)—formerly the Canadian Meteorological Service—and the
Canada Department of Agriculture (CDA). To provide data for regions not
adequately covered by the network, simple regression-type models were developed for computing long-term average soil temperature from long-term air
temperatures (Baier & Russelo, 1970). Subsequently, more complex models
for estimating normal soil temperatures from several climatic variables have
been completed (C. E. Ouellet, personal communication, Ottawa, 1972).

    Climatic observing stations, however, are commonly too far apart to
provide sufficient information. To overcome the lack of atmospheric temperature values, particularly in the northern Great Plains regions, monthly
normals of maximum, minimum, and mean temperatures for any point in
the Canadian Great Plains were estimated from the latitude, longitude, and
elevation of the location using a computer program developed by Williams
and Sharp (1967).

    The normal potential evapotranspiration (PE) was computed from the
estimated maximum and minimum air temperatures by a technique developed
by Baier and Robertson (1965) and programmed by Baier and Russelo
(1968). A sine wave interpolation procedure was used to compute daily
normals of maximum and minimum temperatures from the estimated
normals.

    From these daily estimates and earlier developed techniques (Robertson
& Russelo, 1968; Sly et al., 1971) for computing the normal dates of the first
and last frost dates from temperature normals and other input variables, the
dates of the various seasonal periods, length of period, and degree-days were
computed. This additional information was particularly useful for determining boundaries of zones and map units on the Great Plains, when details were
required and the long-term station records were inadequate to depict the
spatial distribution of temperatures.

Table 2. Descriptive temperature terminology associated with mean tem-
        perature and degree-day ranges.

Descriptive term	Assigned temperature ranges for mean temperatures, °C	Degree days for growing season (>5C)	Code
Extremely cold	<-7	<56	1
Very cold	-7 to <0C	>56 to <555	2
Cold	0 to <5C	<1,110	3
Moderately cold	5 to <8C	<1,221	4
Cool	5 to <8C	<1,388	4
Moderately cool	5 to <8C	<1,665	4
Mild	8 to <15C	<2,220	5
Moderately warm	15 to <18C	<2,775	6
Warm	18 to <22C	>2,775	7
Very warm	22 to <30C		8
Hot	>30		9

Table 3. Descriptive time-scale terminology.

Descriptive term	Assigned period in days	Assigned code
None	0	0
Insignficant	0 to 14	1
Very short	15 to 59	2
Short	60 to 119	3
Moderately short	120 to 179	4
Moderately long	180 to 240	5
Long	240 to 299	6
Very long	300 to 344	7
Nearly continuous	345 to 364	8
Continuous	365	9

Definitions of Terminology

When characterizing conditions within a region, different persons sometimes use similar descriptions to describe different temperature ranges. To enable the use of standardized terminology for characterizing temperature regimes, descriptive terms for ranges in mean temperatures and degree-days were defined (Table 2).

To characterize the length of a season or period, descriptive terms were defined, coded, and associated with given lengths of periods in days (Table 3).

Application

The thermal climate of a particular region may be described by the length of selected periods of the years in relation to its temperature. If the soil temperature for a season was

(1) $\geq 5C$ but $<8C$ for 150 days during the year, the region would be described as having a moderately short, cool growing season and could be coded as 44 P_3 to describe the length of period, the soil temperature range, and the season.

(2) $\geq -7C$ but $<0C$ for 270 days, the region would be described as having a long, very cold, frozen, dormant period and coded as 62 P_1.

The annual climatic condition also could be described or characterized separately for each season (S_1 and S_2) and for each seasonal period (P_1 to P_4) to provide greater detail of the region's thermal characteristics. Where desired, the air temperature may be similarly described.

SOIL WATER CHARACTERIZATION ASPECT OF SOIL CLIMATE

Two parameters, somewhat interrelated, were selected to describe soil water characteristics of a region: (i) water deficient (irrigation requirement) and (ii) climatic moisture index.

Water Deficit

This is the amount of water needed by a crop during the growing season
(≥5C) to meet consumptive use requirements for water not readily available
from the soil. A 5-cm storage capacity and a consumptive use factor of 0.75
PE was used to calculate the water deficit with a 50% risk. If detailed water

Table 4. Illustration of computer program print-out (B133) showing calculated mean and probable soil moisture contents, precipitation, PE,
deficits, and surplus in centimeters at Swift Current, Sask., 1931–60

Variable	Storage, cm	Lowest observed	Percent probability					Highest observed	Mean
			10	25	50	75	90		
Spring									
Soil moisture	2.5	0.0	0.0	0.0	0.3	1.2	2.0	2.5	0.6
	5.1	0.0	0.0	0.8	2.0	3.1	4.1	4.9	1.9
	7.6	0.0	0.0	1.6	3.3	5.1	6.7	7.4	3.4
	10.2	0.0	0.0	2.1	4.3	6.6	8.6	9.0	4.3
	12.7	0.0	0.0	2.2	4.7	7.3	9.7	11.3	4.8
	15.2	0.0	0.0	2.2	4.9	7.7	10.1	11.9	5.0
Growing Season									
Deficit	2.5	9.5	17.6	22.2	27.2	32.2	36.8	39.1	27.0
	5.1	3.8	13.5	18.6	24.2	29.8	34.9	39.1	24.0
	7.6	1.3	10.7	16.2	22.2	28.3	33.8	39.1	22.0
	10.2	1.0	8.9	14.7	21.0	27.3	33.0	39.1	20.8
	12.7	1.0	7.8	13.8	20.3	26.9	32.9	39.1	20.1
	15.2	1.0	7.2	13.3	20.0	26.7	32.8	39.1	19.8
Precipitation		12.5	15.1	19.4	24.1	28.8	33.0	42.5	23.9
PE		55.9	58.5	60.9	63.6	66.2	68.6	70.5	63.5
Surplus	2.5	0.0	0.0	0.8	3.3	5.8	8.1	11.6	3.4
	5.1	0.0	0.0	0.0	0.0	2.1	5.7	8.9	1.3
	7.6	0.0	0.0	0.0	0.0	0.3	3.6	6.4	0.7
	10.2	0.0	0.0	0.0	0.0	0.0	1.9	3.9	0.3
	12.7	0.0	0.0	0.0	0.0	0.0	0.4	1.6	0.1
	15.2	0.0	0.0	0.0	0.0	0.0	0.0	0.0	0.0
Dormant Season									
Precipitation		3.0	4.2	6.3	8.6	10.9	12.9	15.2	9.1
PE		7.7	7.5	8.5	9.5	10.5	11.4	13.5	10.1
Surplus	2.5	0.0	1.3	3.1	5.1	7.1	8.9	10.9	5.1
	5.1	0.0	0.0	0.6	2.7	5.0	7.0	8.6	3.0
	7.6	0.0	0.0	0.0	0.0	2.6	5.0	8.6	1.4
	10.2	0.0	0.0	0.0	0.0	0.0	3.1	8.0	0.6
	12.7	0.0	0.0	0.0	0.0	0.0	0.0	5.4	0.0
	15.2	0.0	0.0	0.0	0.0	0.0	0.0	2.9	0.0
Freeze Season									
Precipitation		0.0	0.0	1.5	4.3	7.2	9.7	13.1	4.5
PE		1.4	1.6	2.2	2.7	3.3	3.9	4.5	2.9
Surplus	2.5	0.0	0.0	0.4	3.0	5.6	8.0	10.9	3.3
	5.1	0.0	0.0	0.0	1.1	3.6	5.8	8.3	1.9
	7.6	0.0	0.0	0.0	0.0	0.9	3.3	5.8	0.7
	10.2	0.0	0.0	0.0	0.0	0.0	0.4	3.2	0.2
	12.7	0.0	0.0	0.0	0.0	0.0	0.0	0.7	0.0
	15.2	0.0	0.0	0.0	0.0	0.0	0.0	0.0	0.0

characterization on a subregional basis is required, the storage capacities can be adjusted to correspond to water-holding capacities of soils of various textures (sands, loams, clay loams) and to landform features affecting drainage.

The water deficit was originally calculated from daily observed precipitation and estimated evapotranspiration by a climatological water budgeting technique as described by Baier and Russelo (1968). The results of the daily computations for various storage capacities, consumptive use factors, and risk levels have been published for some 59 stations across Canada (Coligado, Baier, & Sly, 1968). An illustration of the print-out from the specially developed computer program is given in Table 4.

Monthly precipitation normals were available from publications by the Atmospheric Environment Service. Potential evapotranspiration was computed from standard climatic data by regression-type models developed earlier by Baier and Robertson (1965) and subsequently documented by Baier et al. (1972). The simplest of these models, which require as input only maximum and minimum temperatures and tabulated solar energy at the top of the atmosphere, provided realistic PE estimates that were closely related to Penman's potential evaporation values in various climates (Baier, 1971).

Through the use of "short-cut" regression-type equations developed from the data obtained from the daily computations (Sly & Coligado, in press), it was possible to compute seasonal water deficits for some 700 stations for which monthly temperature and precipitation normals were available. The following relationship gave values for the water deficit within the degree of accuracy considered satisfactory:

$$\text{Deficit (cm)} = 2.01 + 0.259 \ (\text{PE-P})_s + 0.0075 \ (\text{PE-P})_s^2 \qquad [1]$$

where $(\text{PE-P})_s$ is the seasonal difference in the normal potential evapotranspiration (PE) and the normal precipitation (P).

Thus, the soil water condition of a region may be defined broadly for a particular seasonal period covering overall range of textural classes or defined narrowly for a particular textural/topographic situation in terms of surplus (accumulated amount over the capacity) or of deficits (accumulated amount of water required for optimum evapotranspiration).

Climatic Moisture Index (CMI)

The CMI expresses the seasonal precipitation as a percentage of the potential water use by annual crops growing under no moisture stress condition (Sly, 1970; Sly & Baier, 1971).

$$\text{CMI} = \frac{P_s}{P_s + \text{SM} + \text{IR}_s} \times 100 \qquad [2]$$

where

P_S = observed seasonal precipitation

SM = soil moisture storage at the beginning of the growing season as defined for a particular soil/climate situation

IR_S = irrigation requirement or water deficit for the season computed either from daily climatic data or from equation [1].

As a first approach to calculating realistic index values for Canada, it was assumed that

1) The growing season extends from May to September

2) The average consumptive use factor for annual crops during the growing season is 0.75 PE

3) The total amount of water readily available to crops in the soil profile at the beginning of the growing season (SM) is at the maximum which is about 5 cm.

Definition of Moisture Criteria Terminology

To provide uniform description of a soil water condition for any particular seasonal period or site, definitions were assigned to the various ranges of water deficits and climatic moisture indices in Canada (Table 5).

Application

The prevailing moisture regime of a location may be classified according to Canadian criteria, which are based on estimated water deficit and CMI, or may be described according to FAO/UNESCO criteria.

For example, if the seasonal water deficit is 18.1 cm and/or the CMI is 50 (e.g., Saskatoon), the moisture regime would be classified as semiarid, Subclass g (See Moisture Subclasses in the next section). According to FAO/ UNESCO criteria, the moisture regime of this location would be described as "Soil dry in some parts, when soil temperature is >5C, in some years". If the water deficit is 5.8 cm and/or the CMI is 78 (e.g., Ottawa), the moisture regime would be classified as humid, Subclass e. The FAO/UNESCO de-

Table 5. Descriptive moisture terminology associated with ranges in water deficit and climatic moisture index.

Descriptive term	Water deficit range (irrigation requirement)	Climatic moisture index
	cm	
Perhumid	0.0 to <2.5	<84
Humid	2.5 to <6.4	74 to 84
Subhumid	6.4 to <12.7	59 to 73
Semiarid	12.7 to <19.1	46 to 58
Subarid (Boreal)	19.1 to <38.1	25 to 45
(Mesic)	19.1 to <50.8	25 to 45
Arid (Boreal)	≥38.1	<25
(Mesic)	≥50.8	<25
Xeric		<25

scription would read "Soil not dry in any part as long as 90 consecutive days in most years".

Thus the moisture characteristics of the soil in moist unsaturated regimes can be classified according to Canadian criteria, if estimates of the seasonal water deficit and of the CMI are available. Otherwise, descriptive terminology would be used following guidelines given in the next section under Moisture Subclasses.

PRIMARY CLASSIFIERS AND GENERAL CHARACTERISTICS OF THE SOIL CLIMATE CLASSES AND SUBCLASSES

Temperature Classes

1. ARCTIC

[3]Mean annual soil temperature: $<-7C$
[4]Growing season ($\geqslant$5C): Not significant ($<$15 days)
Thermal period ($\geqslant$15C): None
Continuous permafrost at $>$50-cm depth

1.1 Extremely Cold Arctic
    [4]Dormant season ($<$5C)—dominant

2. SUBARCTIC

[3]Mean annual soil temperature: -7 to 2C
[3]Mean summer soil temperature: 5 to 8C
[4]Growing season ($\geqslant$5C): Moderately cold (5 to 8C)
                                Degree days $<$555
                                Short ($<$420 days)
Thermal period ($\geqslant$15C): None
(Discontinuous permafrost may occur at $>$50 cm depth. Alpine soils are included in this group).

2.1 Very Cold Subarctic
    [4]Growing season ($\geqslant$5C): Degree days $<$555

3. CRYOBOREAL

[3]Mean annual soil temperature: 2 to 8C
[3]Mean summer soil temperature: 8 to 15C
[4]Growing season ($\geqslant$5C): Mild (8 to 15C)
                                [4]Degree days $<$1,250
                                Length 120 to 220 days

---

[3]Primary classifier for class and interphases with FAO/UNESCO Soil Climate Map of North America.
[4]Subclassifier for Soil Climate Map of Canada.

Thermal period (≥15C): Not significant
                        Degree days <33
[Undisturbed soils are usually frozen in some part of control section
(<1 m) within some part of dormant season].

3.1 Cold Cryoboreal
Growing season (≥5C): [4]Degree days <1,110
                        Moderately short (<180 days)
                        Cool (2 to 5C)

3.2 Moderately Cold Cryoboreal
Growing season (≥5C): [4]Degree days <1,221
                        Moderately long (<220 days)
                        Moderately cool (5 to 8C)

4. BOREAL

[3]Mean annual summer temperature: 5 to 8C
[3]Mean summer soil temperature: 15 to 18C
[4]Growing season (≥5C): Moderately warm (15 to 18C)
                          Degree days 1,250 to 1,720
                          Moderately short to moderately long (170 to
                              220 days)
Thermal period (≥15C): Moderately warm (15 to 18C)
                        Degree days 33 to 222
                        Short (<120 days)

4.1 Cool Boreal
[4]Growing season (≥5C): Cool
                          [4]Degree days <1,388
                          Moderately short >170 days
Thermal period (≥15C): Very short (<60 days)

4.2 Moderately Cool Boreal
[4]Growing season (≥5C): Cool
                          Degree days <1,720
                          Moderately long (<220 days)
Thermal period (≥15C): Degree days <220
                        Short (<120 days)

5. MESIC

[3]Mean annual soil temperature: 8 to 15C
[3]Mean summer soil temperature: 15 to 22C
Growing season (≥5C): Moderately warm to warm (15 to 22C)
                        Degree days 1,720 to 2,775
                        Moderately long to continuous (200 to 365
                            days)
Thermal period (≥15C): Degree days 167 to 666
                        Short to moderately short (90 to 180 days)

5.1 Mild Mesic

  Growing season (⩾5C): Mild
                       Degree days <2,220
                       Moderately long 200 to 240 days
  Thermal period (⩾15C): Short <120 days

5.2 Moderately Warm Mesic

  Growing season (⩾5C): Moderately warm
                       Degree days <2,775
                       Moderately long to continuous (>240 days)
  Thermal period (⩾15C): Moderately short (<180 days)

6. THERMIC

  [3]Mean annual soil temperature: 15 to 22C
                                   Moderately warm to warm

7. HYPERTHERMIC

  [3]Mean annual soil temperature: >22C
                                   Very warm to hot
  (Thermic and hyperthermic regimes do not occur in Canada).

Examples for comparable soil temperature classifications at selected locations in the USA using Canadian and FAO/UNESCO criteria are given in Table 6. The computer program (B121) specially developed for determining

Table 6. Comparison of computed growing season and mean annual soil temperatures and their assigned codes at selected locations in the USA.

Station	From	To	Growing season ≥5° C				Annual soil temperature		Map[†] code
			No. of days	Mean temp.	Degree-days	Code*	Mean	Code*	
				°C	°C		°C		
Bushland, Texas	1	365	365	14.8	3,575	≥6	15.0	5 or 6	5f or 6f
Archer, Wyoming	90	332	242	13.7	2,093	5	9.8	5	5h
Lincoln, Nebraska	93	339	246	16.4	2,805	6	11.4	5	5e
Huntley, Montana	95	333	238	13.9	2,110	5	9.8	5	5h
Tempe, Arizona	1	365	365	19.3	5,199	7	19.3	6	6j or 7j
Lexington, Kentucky	67	365	298	15.6	3,146	≥6	13.3	5	5e

* Taken from Section IV-A. No exact degree-day ranges available for temperature classes 6 and 7.
† Abstracted from Soil Climate Map of North America as given in: Soils of Canada, Research Branch, Canada Department of Agriculture, Ottawa.

soil temperature criteria of the growing season for Canada was used with USA monthly "normals". These were generated from monthly soil temperature data at 50-cm depth as provided by F. Newhall (Personal communications, Washington, D. C. 1971) and averaged over 4 to 5 years.

Moisture Subclasses

1. AQUIC REGIMES

Soil is saturated for significant periods of the growing season.

> a. *Peraquic*
> Soil saturated for very long periods
> Ground water level at or within capillary reach of the surface
> b. *Aquic*
> Soil saturated for moderately long periods
> c. *Subaquic*
> Soil saturated for short periods

2. MOIST UNSATURATED REGIMES

Varying periods and intensities of water deficits during the growing season.

> d. *Perhumid*
> [5] Soil moist all year, seldom dry
> [6] No significant water deficits in the growing season
> Water deficits 0 to <2.5 cm
> CMI >84
> e. *Humid*
> [5] Soil not dry in any part as long as 90 consecutive days in most years
> [6] Very slight deficits in the growing season
> Water deficits 2.5 to <6.4 cm
> CMI 74 to 84
> f. *Subhumid*
> [5] Soil dry in some parts when soil temperature is ⩾5C in some years
> [6] Significant deficits within the growing season
> Water deficits 6.4 to <12.7 cm
> CMI 59 to 73
> g. *Semiarid*
> [5] Soil dry in some parts when soil temperature is ⩾5C in most years
> [6] Moderately severe deficits in growing season
> Water deficits 12.7 to <19.1 cm
> CMI 46 to 58

---

[5] FAO/UNESCO criteria.
[6] Canadian criteria (calculated).

h. *Subarid*

[5] Soil dry in some parts or all parts most of the time when the soil temperature is ⩾5C

Some periods as long as 90 consecutive days when the soil is moist

[6] Severe growing season deficits

Water deficits 19.1 to <38.1 cm in Boreal and Cyroboreal regimes, 19.1 to <50.8 cm in Mesic or warmer regimes

CMI 25 to 45

j. *Arid*

[5] Soil dry in some or all parts most of the time when soil is ⩾5C

No period as long as 90 consecutive days when soil is moist

[6] Very severe growing season deficits

Water deficits ⩾38.1 cm in Boreal regimes and ⩾50.8 cm in Mesic or warmer regimes

CMI <25

x. *Xeric*

[5] Soil dry in all parts 45 consecutive days or more within the 4-month period (July to October) following the summer solstice in more than 6 years out of 10.

[5] Soil moist in all parts for 45 consecutive days or more within the 4-month period (January to April) following the winter solstice in more than 6 years out of 10.

(Xeric regimes are not believed to occur extensively in Canada but may be found in local areas of microclimate).

Computations of soil water characteristics and corresponding index codes using Canadian criteria as compared with the map coding using FAO/ UNESCO criteria for a number of locations in the USA are presented in Table 7.

SOIL CLIMATE MAP OF CANADA

The described criteria for the soil temperature and soil water aspects were used in the preparation of the Soil Climate Map of Canada together with other relevant soil and climatic information including the map of permafrost by Brown (1967) and the atlas of Climates of Canada for Agriculture by Chapman and Brown (1966). Relationships of vegetational associations with climatic soil areas, the report and map of the Forest Regions of Canada (Rowe, 1959), and studies by Coupland (1961) on the grassland classification in the northern Great Plains were also considered. Boundaries of the map zones followed known physiographic features of the landscape together with soil survey results. The map is of a scale of 1:5,000,000 on the Lambert Conformal Conic Projection so that it can be easily related to the map unit areas of the Soil Map of Canada. A colored version of the Soil Climate Map is being published by the Research Branch, Canada Department of Agriculture, Ottawa. Two small scale simplified maps (black & white) are

Table 7. Computation of soil water characteristics and corresponding codes at selected locations in the USA.

Location	Latitude N ° '	Longitude W ° '	Seasonal precipitation cm	Seasonal PE cm	Water deficit IR$_2$,7.5 cm	CMI index	Index code*	Map code†	Index error‡
Byers, Colo.	39 42	104 13	21.6	88.0	52.2	27	h	5h	
Sterling, Colo.	40 37	103 12	25.0	84.5	43.9	34	h	5h or 5f	-3
Las Animas, Colo.	38 05	103 13	20.5	99.5	69.2	22	j	5h	-2
Huntley, Mont.	45 55	108 15	15.6	78.1	47.5	23	j	5h or 4e	
Havre City, Mont.	48 34	109 40	20.1	69.7	33.3	34	h	4h	-10
Hay Springs, Nebr.	42 41	102 41	33.4	79.7	30.1	49	g	5f	-4
Holdrege, Nebr.	40 26	99 23	38.1	80.6	26.6	55	g	5f	-12
Valentine, Nebr.	42 52	100 33	30.2	75.7	29.4	47	g	5f	-8
North Platte, Nebr.	41 08	100 42	34.4	78.4	28.0	51	g	5f	-4
Williston, N. Dak.	48 09	103 37	22.3	64.5	26.3	42	h	4g	-2
Devils Lake, N. Dak.	48 07	98 52	30.5	62.3	17.9	57	g	4f	
Edgely Exp. Farm, N. Dak.	46 20	98 42	35.9	69.2	18.9	60	f	4f	-13
Dickinson, N. Dak.	46 47	102 48	26.5	68.4	26.0	46	g	4f	-4
Mandan Exp. St., N. Dak.	46 48	100 54	31.7	67.7	21.1	55	g	5h	
Kenton, Okla.	36 55	102 58	23.0	93.3	57.4	27	h	6f	
Ponca City, Okla.	36 42	97 05	51.3	89.3	22.7	65	f	6f	
Lawton, Okla.	34 36	98 24	48.0	90.6	26.7	60	f	6f	-12
Woodward, Okla.	36 27	99 23	37.2	90.0	36.6	47	g	5h or 5f	
Ardmore, S. Dak.	43 01	103 70	24.3	90.1	51.7	30	h	4f	-22
Newell, S. Dak.	44 40	103 50	23.9	76.2	36.1	37	h	5f	-28
Dalhart, Tex.	36 03	102 50	24.7	88.6	49.3	31	h	5e	-1
Platteville, Wis.	42 54	90 50	49.3	74.8	13.5	73	f	5e	-2
Hays, Kan.	38 52	99 20	41.6	83.4	26.0	57	g	5f	-14
Greensburg, Kan.	37 36	99 18	35.6	91.0	39.4	45	h	6f or 5f	-11
Hoxie, Kan.	39 21	100 27	36.5	87.3	34.6	48	g	5f	

* Climatic moisture index code (See Section IV-B): d = >84, e = 74-84, f = 59-73, g = 46-58, h = 25-45, and j = <25.
† Taken from Soil Climate Map of North America as given in Soils of Canada, Research Branch, Canada Department of Agriculture (In press).
‡ Deviation of computed moisture index from lowest value of subclass taken from map.

Table 8. Extent of soil climatic classes and subclasses in Canada (100 km²).

Class and subclass	Arctic	Subarctic	Cryoboreal	Boreal	Mesic	Subclass total	% of total
			—100 km²—				
d. Perhumid	--	7,216	10,782	4,141	250	22,388	25.8
e. Humid	26,180	19,577	10,077	606	911	57,352	66.2
f. Subhumid	--	546	3,524	313	62	4,447	5.2
g. Semiarid	--	--	409	1,181	5	1,618	1.8
h. Subarid	--	--	--	914	--	914	1.0
Class total	26,180	27,340	24,794	7,180	1,227	86,720	
Percent	30.2	31.5	28.6	8.3	1.4	100	

presented in Fig. 2 and 3 to illustrate separately the soil temperature classes and the soil moisture subclasses for Canada.

SOIL CLIMATIC REGIONS IN CANADA

The scheme for characterizing the soil environmental climatic conditions illustrates directly the general climatic conditions under which the soil was formed, the climatic potential for biological activity, and the limitations existing for the use of the soil resource. The extent of each zonal climatic class and subclass as obtained from the 1:5,000,000 Soil Climate Map of Canada is presented in Table 8.

Over 2,618,000 km² or 30.2% of Canada has an Arctic climate where biological activity may occur only briefly once every 5 years. Another 2,734,000 km² or 31.5% has a subarctic climate which also severely restricts biological activity. Thus out of a total of 86,720,000 km², only 38% have sufficient thermal energy for sustaining productive vegetations and of this amount (Cryoboreal, Boreal, and Mesic) only 34% does not have severe moisture limitations (perhumid, humid, and subhumid). However, much of the Cryoboreals are suitable for the production of small grains and forages even though they are not productive for other crops.

ACKNOWLEDGMENTS

J. S. Clayton of the Soil Research Institute, Research Branch, Canada Department of Agriculture, has been actively involved in developing the classification system and the Soil Climate Map of Canada. Members of the Agrometeorology Section, Plant Research Institute, especially W. K. Sly, contributed significantly to this study by developing techniques and providing derived data. The data processing staff of the Agrometeorology Section wrote the computer programs and arranged for all computations and plottings through the Data Processing Division of the Canada Department of Agriculture.

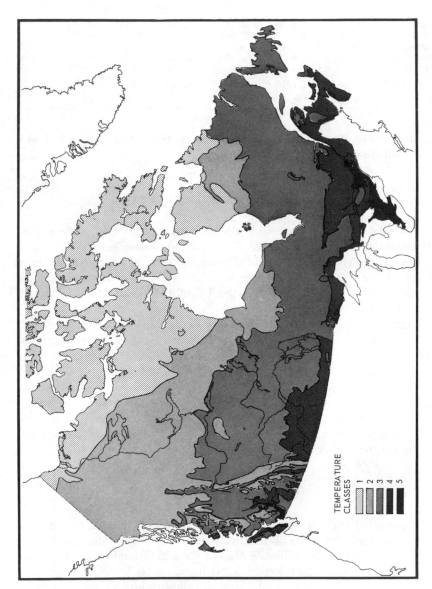

Figure 2. Sample map of soil temperature classes abstracted from the Soil Climate Map of Canada.

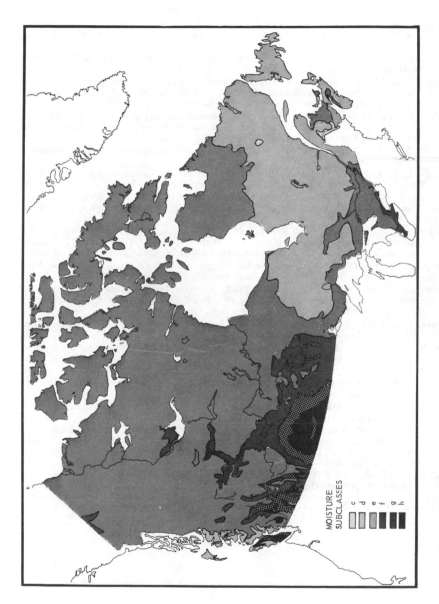

MOISTURE SUBCLASSES

c
d
e
f
g
h

Figure 3. Sample map of soil moisture subclasses abstracted from the Soil Climate Map of Canada.

LITERATURE CITED

Baier, W. 1971. Evaluation of latent evaporation estimates and their conversion to potential evaporation. Can. J. Plant Sci. 51:255–266.

Baier, W., D. Chaput, D. Russelo, and R. Sharp. 1972. Soil moisture estimator program system. Tech. Bull. No. 78. Agrometoerological Section, Plant Research Inst., Can. Dep. Agr., Ottawa.

Baier, W., and G. W. Robertson. 1965. Estimation of latent evaporation from simple weather observations. Can. J. Plant Sci. 45:276–284.

Baier, W., and D. A. Russelo. 1968. A computer program for estimating risks of irrigation requirements from climatic data. Tech. Bull. No. 59. Agrometeorology Section, Plant Research Inst., Can. Dep. Agr., Ottawa. 48 p.

Baier, W., and D. A. Russelo. 1970. Soil temperature and soil moisture regimes in Canada. p. 35–65. *In* Proc. Canada Soil Survey Comm., 8th Meeting (Central Exp. Farm, Ottawa, October 1970).

Brown, R. J. E. 1967. Permafrost in Canada. Map and explanatory notes. Geological Survey of Canada. Map 1246A.

Chapman, L. J., and D. M. Brown. 1966. The climates of Canada for agriculture. Rep. No. 3. Canada Land Inventory, A.R.D.A. Dep. Forestry Rural Develop., Ottawa.

Clayton, J. S. 1971. The soil climates of Canada. p. 21–26. *In* Proc. Nat. Tech. Work-Planning Conf. Co-op. Soil Survey. USDA, Soil Conservation Service, Charleston, S. C.

Coligado, M. C., W. Baier, and W. K. Sly. 1968. Risk analyses of weekly climatic data for agricultural and irrigation planning. Tech. Bull. No. 17–58, 61–77. Agrometeorology Section, Plant Research Inst., Can. Dep. Agr., Ottawa. 8 p., 26 tables.

Coupland, R. T. 1961. A reconsideration of grassland classification in the Northern Great Plains of North America. J. Ecol. 49:135–167.

Robertson, G. W., and D. A. Russelo. 1968. Freezing temperatures risk calculations: systems analysis and computer program. Tech. Bull. No. 60. Agrometeorology Section, Plant Research Inst., Can. Dep. Agr., Ottawa. 31 p.

Rowe, J. S. 1959. Forest regions of Canada. Forestry Branch Bull. 123. Dep. Northern Affairs Nat. Resour., Ottawa.

Sly, W. K. 1970. A climatic moisture index for land and soil classification for Canada. Can. J. Soil Sci. 50:291–301.

Sly, W. K., and M. C. Coligado. 1973. Agroclimatic maps for Canada—derived data; Moisture and temperature regimes. Tech. Bull. No. 81. Agrometeorology Section, Plant Research Inst. (in press).

Sly, W. K., and W. Baier. 1971. Growing seasons and the climatic moisture index. Can. J. Soil Sci. 51:329–337.

Sly, W. K., Geo. W. Robertson, and M. C. Coligado. 1971. Estimation of probable dates of temperatures near freezing from monthly normals, station elevation, and astronomical data. Tech. Bull. No. 79. Agrometeorology Section, Plant Research Inst., Can. Dep. Agr. 21 p.

Williams, G. D. V., and W. R. Sharp. 1967. A program to estimate normals of temperature and related agroclimatic elements for locations on the Canadian Great Plains. Tech. Bull. No. 11. Agrometeorology Section, Plant Research Inst., Can. Dep. Agr., Ottawa. 27 p.